NEOGENERATION NEOEVOLUTION

THE NATURE OF LIFE IS TO A CHANGE OF NATURE,
IT IS MORE THAN EMC2, IT'S...AI > DNA > M.

PAWEL KOZYCZ

Contents

Contents

Contents

Contents

Contents

Foreword

Is it possible to exceed physical, biological, and economic parameters millions of times and practically indefinitely?

What mechanism is needed to achieve such effects? Does it have to be a miracle?

So, can one create something out of nothing?

Do we have to be God or, to put it differently, some naturally intelligent system - NI (natural intelligence), i.e., replacing artificial intelligence, in creating our neo-evolutionary conditions, changing the current parameters of the broadly understood DNA of physics, biology of the micro-macrocosm, first by simply leaving the passive consumer attitude towards more active administration of DAA->AI infinite process resources, which have this invisible advantage over the material primitive approach to phenomena, where these process, this nothing, becomes a base for creating from this process nothing, the factual basis for participation in this non-passive creation of laws despite the previous one, even the most mathematically confirmed on the body of previous actions, understanding, perception?

Answer. Neogeneration of space/nature <- neo infrastructure <- economy = responsibility -> laws, i.e., NI->DNA<-DAA-AI of the matter of life...

Pawel Kozycz - author of previous futuristic, new age, new generation or better to say a neo generation, neoevolution, neo science, neo religion books because of neoview on nature/Got ; " Masters of Life and Universe", "100 Verses for Life and space re-creation", "Neo Creation of Life and Space/Matter or Bricks of Micro Drone Revolution", " Towards the foundation of the mechanism of neo-creation", "Beyond the nature of life and stars", " Neo evolutionary blessing", " A handbook of re-neo processing of life and matter", "Matter is process", etc.

CHAPTER ONE

Joining the neo/evolutionary expansionary domino effect of the mechanism/model of the universe of matter and life.

DAA as ... a political tool against evolutionary, economic, and political tensions.

A luring mechanisms, factors – "x-factors", and their production, adding to the existing processes and structures known to us ...

The arising of something out of "nothing" ... Such phenomena do occur (physics of elementary particles).

That is, a specific initiation of phenomena, processes, structures from completely unknown areas, such anonymous experimentally or pregenetic (DAA) initiatives in which we can be more or less active/passive participants.

But every more or less known system for its part is vying for its position. It may be a system that we can; we are its live suppliers, which decides how long, how we have to exist for it, but ... roles can change, we can change.

It is dependent on this expansionary domino model in taking initiatives, decisions, changing, evolving structures, processes in which we and this, these systems participate ...

It is a matter of exchange, trade (with the environment, its systems, the production of one's own systems and niches), production, mega-production or hyper-evolution, overtaking, applying the hitherto state of evolutionary relations. Create a system that is more and more efficient.

The world will become more with its potential; we will be our mega-evolutionary initiatives to expand our possibilities of existence and development.

We will not feel like rats in a cramped cage in which these rats feel more aggressive towards each other because of this ...

We are talking here about the relations between humans, inter-environmental in our planet's environment at a given level of development, which until today does not provide sufficient conditions to increase the guarantee of life, not to mention development ...

the hands of the earth's 24-hour existential clock are set to 2 minutes before midnight ... and this is mainly due to fatalistic politics, or rather the lack of human politics on this earth ...

And all this is due to a peculiar invisible hand of the market, a complex man in a neo-evolutionary way.

There is a need to indicate a more precise control mechanism responsible for human environmental and infrastructural processes.

Delegating this enchanted hand to make decisions within the framework of the neo-evolutionary decision-making mechanism, at every level of his economic activity, i.e., economic, i.e. corresponding to the mutual improvement of mutual environmental relations, trade exchange, i.e. hyper-environmental, hyper-ecological, hyper-economic investments, which are to be the backbone of existence, i.e., the economy, politics on a local, global and micro-macro level of more or less animated matter.

Yes, it is a political decision, i.e., taking responsibility for the environment (finding its rights ... = establishing its laws that we create it, not so much realizing that through this we influence these laws, developing new technologies/mechanisms, mega / production for them = hyper evolution of the laws/=rules and mechanisms/tools of this environment ...), in which live among other people, who should have an increasingly efficient decision-making system, having involuntary pregnant effects on global changes.

Each decision in shaping the hyper-evolutionary environment is a political decision ... yes, decisions regarding scientific and economic progress are strictly political decisions and strictly market decisions, strictly economic ones.

The inclusion of a specific application, tools in deciding about our participation in the neo-evolutionary process. Our share that is our contribution, our tax, that is, our literally developed law, fairly calculable commercial law in directing funds to goals that are

fundamental for the directions, chances of existence, and development of each of us together and separately.

Such a tax mentioned 10% anonymous share/vote at every level of participation in the human market, as well as the discovery/ development of this share in every structure, process of the environment/infrastructure, which would be a litmus test and also a dynamic driver (DAA), such a super evolutionary invisible hand/ decision. Because it will be a tax, the destination of which on the taxpayer's part will be kept secret. His decision/lack of decisions in the directions of transferring his participation will have a powerful total effect on the fate of the environment, in which resources and investments will be redirected, according to deep intuition or according to trust institutions, decisions / and their absence will also be a fateful decision for each of us and environment in

multi expansionary domino effect...

It is them that this tax share, contribution, each next single domino cube, has this magical influence on the directions of development of science, economy, environment, the universe of matter and life in which we are, and in which this sharing tool gives a greater chance to participate more consciously, a greater right to life, for which this participation determines this law of basis and strength.

Such factor x, additional DAA factor for diversification, an indication of the deepness of Esk potency of existing mDNAn environmental processes ... at the evolutionary, economic, environmental, infrastructural, global, local, political, and scientific levels.

By introducing DAA at every decision-making level, including automatic / production / scientific

Eskm (DNA) DAAn – creating platforms above, beyond the current contradictory interests, phenomena, patterns, models ...

Blessing of support of divine enterprises for new hyper environmental platforms of life and matter...

means DAA(aware evolutionary decision)x m(infrastructural/ environmental niche) = Esk(direction efficiency).

So we have come to the next stage of implementing, finding, and applying natural and artificial niches to stimulate and control the evolution of matter and life processes.

I do not impose but still inspire; I propose applying systems, opportunities that today and in a dynamic time of changes, trends, and potentials, it would be possible to control, participate and initiate more and more freely.

It is necessary to find and work out this constant coefficient of economic control of processes by a man and systems that are more or less dependent naturally and unnaturally – production, processing, operational, and IT systems.

The so-called 10% private, public, systemic industry relief for pregenetic, pre-ecological, and pre-economic neo-evolutionary investments is the key to balanced control according to the interests of each of the interested parties, including biological, physical, production, economic, research and, above all, developmental algorithmic systems.

We are talking about investments in the production and use of natural, artificial, environmental, and infrastructural niches on the macro and micro scale, broadly understood matter and life.

These hyper-evolutionary investments, according to this anonymous/purely{a sort of neoevolutionary/creative blessing!} intellectual pre-genetic factor (such a specific factor from nowhere, dark energy/hyperevolutionary matter), a dynamic factor influencing the directions of material/decision support as if from nowhere, but through more or less autonomous systems that are part of the whole an algorithmic system influencing the pace, force of change at the level of activity, which could qualitatively, quantitatively contribute to the level of security and development of an activity, where a man would be the last conscious decision-maker, a link giving certain competences [... blessing competencies] to the automatic economic, production, ecological systems – economically, politically, politically over the particular interests of this hyperevolutionary machine.

Decision-making anonymity, as well as the voluntary transfer of one's production – (=) neo-evolutionary

production competencies to trust systems, is supposed to make human decision-making independent, as well as ...replacing the decisions of the broadly understood DNA systems of the environment of matter and the biology of the micro-macro cosmos, with the help of more and more own

(less and less dependence on other systems of the nature of the environment, but also on its own infrastructure that can give the direction of any existence, development in infinite potential)

algorithmic systems, trust, information and control systems, in purely economic structures, as well as purely structures of matter and life, structures, micro-processes, macrocosmic systems – which are also guided by their economies – often influencing our existential and development perspectives without our participation, participatory ... perceptual ...changing the perception of the evolution of matter and life in systems that seem not to be dependent on us, us addicted to them ... but reciprocally, where the strength of a given side of these systems depends on the use of the infinite the resource of the potential of matter and life, and above all the process-factor DAA of the decision in this exploration, the exploitation of this infinite potential.

Predictive decision-making building blocks are neo / evolutionary processing of matter and life ... through new hyperevolutionary systems, forces, inputs.

The issue of autonomous accelerates the process that occurs, tries to demonstrate and at the same time better apply in this civilization activity, i.e., hyper neo-evolutionary, in fact, civilization, i.e., aware of its active and not only passive role in shaping, in accelerating participation in the competition of systems, in shaping structures and their properties in the existential and developmental process, in the infinite potential of threats and the chances of the cosmos in us and beyond.

It is about generating, pre-generally self-propulsion according to the appropriate more or less direct algorithmization of directions of self-developing natural and artificial platforms through non-final but flexible decision-making systems of human intervention.

10% system through this autonomous scientific, economic, hyper ecological, hyper evolutionary stabilizer ... going beyond any closed systems, systems, relations, and models, e.g., Emc2,

based on the principle, thesis, the law of the infinite potential of the niches of matter and life of the micro-macro cosmos, at every particular level and direction of structures, processes of matter and life, not always / never completely unknown, used ... and their level of perception, the exploitation of the potential depends on this anonymous, by no means a decisive factor pushing further this whole system, the cosmos, i.e., the DNA decision-making factor, or rather the pre-DNA, i.e., DAA – dynamic administrative activity, administration of the entire cosmos-life.

CHAPTER TWO

Pre-regeneration of structures, processes of shaping the properties of matter, and the life of micro and macro cosmos in us and beyond.

We are entering a new stage of our considerations, searches, conclusions, formulating theses in shaping, processing, and developing models of environmental transformation through hyper-industrialization, expanding possibly mega-production, automatic, autonomous, quasi-autonomous natural, neo-neutral background,

defining its own DAA and not the environmental DNA status quo of the micro-macro cosmos ...

Specific planning of structural and, therefore, property transformations. The property changes of the entire, partial area, space of the environment, the composition of which, the active component of which will be contributed to our systems, and not only others ...

Without praying to the tree / natures systems, but blessing oneself, – this tree, DNA systems have to pray to us, that is, be subject to our DAA, and then DBA – DAA (b) DCA – DAA (c), etc. transformations ... where this transformed micro-macro-cosmos system, partly its internal and external, will give us further clues, arguments, strength, old passive DNA, and further a new more active DAA blessing in shaping further perspectives of the entire micro and macro cosmos already only to the absolute degree, i.e. the absolute model DAA = m / Esk ... as we say about the absolute properties of the Emc2 formula, in this case we are talking about more profound, more universal, economic implications using the EskmDAA model, possibly the most primitive, i.e. more genetic , that is, the pre-genetic model of shaping all properties depending on the level of surveillance, penetration, perception of new, more profound, artificial, i.e. ours system, not only other systems of transformation of the properties and structures of niches ... of matter and life.

We are talking about the go-to-whole method, i.e., hyper ecology, i.e., maximization/automation of infrastructural changes, maximization of space occupation, areas not even at first glance needed in the near or even further time ... a specific occupation, absolute colonization – only limited by certain ecological elements – an increasing area of the potential of the neo-infrastructural environment to the limits of possible, ecological stability, i.e., not leading to greater threats in the future than the benefits of these neo-environmental, infrastructural undertakings, which are a feature of all phenomena, processes in the so-called nature, i.e., the infinite cosmic system which also includes, initiates their trends, tendencies affecting behavior, properties, structures, deeper structures, and further factors influencing properties, seem to be constantly or unrealistic to any change, achievement.

It is a struggle to shift the margin of survival, always involving the dilemma of surrendering to other systems and properties or initiating our own to increase our buffer of safety, existence, and development. A buffer of rights and obligations, or rather obligations resulting from this law, properties more or less enshrined in models of structures, relations, and processes, to which we can artificially refine, also creating neo-infrastructural, neo-ecological, neo-environmental niches. Their constitutions ...- Neo-evolutionary mathematical models ...

Patterns supporting perception, competition with ... ourselves, with our environment, with our ecology, and our economy with our environment/cosmos at an increasingly higher existential, developmental, and civilization level

It is a kind of human constitution with an infinite potential of the environment of an infinite number of systems.

The infinite potential of the systems entering into it, and thus the potential for benefits, the efficiency of which is based on the proper observance of obligations, and behind them, the place in the environment to which this hyperevolutionary / neocreational neo-mathematical constitution has been assigned.

We can have our own link, link, niches, and not only ... in perception, amortization, in an increasingly larger model of the dependence matrix of any phenomenon, process, structure, properties transforming to infinity ... initiative/energy, from the very "thought", or a new niche from DAA, dynamic over environmental, over "natural" decision to m / Esk.

Systems are always closed, but infinite potentials, infinite decks, and factors, including own new initiatives .. transforming the "closed" system makes it final .. for another initiative DAA <DBA <DCA ... which is the beginning, this anonymous infinite pre-genetic factor ... extra infinite factorial ... pre hyper proprietary initiative of the whole, the entire evolution of entire micro-macro cosmos of the infinitive potential of structures.

CHAPTER THREE

We are talking about a 100% plasticity of all processes, a structure of dependencies based on the infinite potential of all relations in terms of interpretation, intervention, ... connection, switching from a specific side, a base of reactions ... also from nowhere.

The issue of selective economics, i.e., economic, host participation in this process of evolution, including, in this case, its own contribution, the infinite potential of these artificial mines interacting with nothing ... potentially and truly infinite own properties into micro macrocosmic transformations.

The issue is proper, non-surrendering, and at the same time responsible for the involvement in shaping the ownership, effectiveness of transformations ... the domino effect – from turning a smaller domino to a larger one – leading to the transformation of inter-infrastructural relations with the environment according to the increasingly dense network of targeted and directional interactions ... a specific domino effect of an increasingly productive, automatic, autonomous influence on further structural and process transformations, infrastructural and environmental properties, faster and faster, in a race with other constant and changing physical, biological and chemical systems of the world ... hidden by whole other systems/civilizations in atoms, stars, dark matter, etc.

It is simply about further, but from the perspective of civilization, work on the improvement, productivity, and economics of working on this process, and above all, the development of the awareness of hyperevolutionary intelligence in this aspect of development and existence.

This shapes the constitutional direction of work, processing, a specific neo-evolutionary, hyperevolutionary, and hyperevolutionary productivity.

We can; we must treat ourselves to the blessing of neo-evolutionary participation, the strategy of striving, thoughts ... to cram a new account, to include the goal in this major, non-major model of decision-making infinity DAA in infinity DBA in infinity DCA ... n .... DAA (a) <DAA (b) <DAA (c) <.... DNA ...... DAA (n) .... DXA ... DAA (x)

What we want to know is what is mine, what is hyper evolutionary

not only addiction, but transforming relations at a given level of intervention ... the interference of a peculiar anonymous voting system of a hyper-economic nature that economizes each environmental relationship, speaking here about selective economics in direct interaction with the environment, and not ignoring it or giving in to it – which is also ignoring it – which is also initiative from nowhere, a specific invisible decision-making spirit inherent in each of us literally and figuratively ... in atoms also ...

Sketching the beginning of the domino effect, which from the invisible force, the self-will of wanting initiative, from the slightest change, can affect chain reactions with an unimaginable effect but always potentially achievable ...

It is the generation of always closed systems that always have a gene in them, sometimes common to other generic influences of an infinite number of systems and potentials ... in which it is possible to work out, and co-

produce the proper directional role in the hyper generic development of the environment of the matter and the life of the micro-macro cosmos in us and beyond.

We are talking about the positioning of our being in the evolutionary processes of the micro-macro cosmos, i.e., about hyper evolution, i.e., a system of constant improvement of the infrastructure for tools transforming the existing environment to surpass the existing infrastructure, i.e., the environment of systems in us and beyond – just parallel – bypass – industrial, economic, evolutionary ecological, cosmic, biological ... park for new life and matter, a new EvmDNA, means no more just EvmDNA / ... Emc2 but EskmDAA, .. EskmDAAn ...

The potential of more and more precise, efficient infrastructure tools for going out to the so-called other worlds that are parallel and also affect the properties of Emc2 with another c, DNA with other molecules ..., our own worlds, also crossing each other, influencing – the question of our activity in a direction ... to be stronger than the cosmos we know, creating its own infrastructural planes, surpassing other planes at a given level of direction. That is what makes our own hyper-artificial sun, our own big bang, and other incubators of new properties of the structures of matter and life processes more efficient.

Big Bang bis and other encores

From breaking the unattended leg bones to their reassembly – building parallel worlds from blocks more minor than the previous ones to shape the direction of our own evolutionary sequences, e.g., quantum determinants, DNA, etc., –

for this, the need for investment in tools and materials increases metabolism's technological and productive level at lower and higher levels.

To be continued with neo constitution of matter and life by artificially Big Bang means by EskmDAAn.

CHAPTER FOUR

Neo-evolutionary technology, production, economics, and science as simply constantly raising the technical level of civilization, social, existential, and methodological relations ... that is, however, technology can soften views, customs, models / scientific frameworks, consolidate the proper shaping of environmental relations, interpersonal and inter-system infrastructures the matter and life of the micro-macro cosmos in the full sense of the word.

Peculiar participation in neo / evolutionary transformations raises and overrides man's position in a dynamic inter-systemic environment.

This image of participation in the infinity of neo / inter / evolutionary perpetual motion potential presents differently the perspectives of man, his science and economy, his level of social aspirations – so social! that is, between the systems – in the mega-cosmic relation no longer cosmic, no longer micro-cosmic but hyper micro macro-cosmic.

We are no longer talking here about a closed area of human activity and its perspectives. Our life will be more reminiscent of resembling more artistry of neo/transformation, neo/creation than some vegetative or rather destructive existentialism.

Such a divine role (God is in us – we must awaken him in ourselves), participation in the infinite tendencies of environments, in their participation.

Collaborative – responsible participation as the basis for developing hyper/neo / inter-evolutionary, neo / inter-cosmic new world, new environment, the birth of one's own initiatives/ influences.

So we are talking about control, organized, co-responsible system, a division in mega-production enterprises, a hyper cosmic machine with a two-sided, multilateral, hyper-ecological, hypernatural risk, shifting the boundaries of the impact on the existential potential, our potential, the potential of our environment, system, and positive potential. Other environmental systems in us and beyond ... the future of systems, worlds, the so-called natural and artificial niches, which are to be, and are the platform of our existence, our perspective, obstacles, breaking further imaginary/illusory boundaries of the development of models of physics and biology, economics, ...

Nature is a machine, or rather a production, mega-production system; its components, tools, mega/evolutionary, mega/ecological components, i.e., by infinite potential.

Furthermore, here we are interested in the control room, this tool, becoming an increasingly more prominent participant, a machine environment, simply an environment/infrastructure, – The most powerful tool for an increasingly appropriate system of use, control, as it is done by other competing systems of matter and life in the micro and macro cosmos.

This hypernatural, hyper ecological, hyper scientific, hyper economic approach has an educational, formative, organizational, and even comfortable religious character in further influences ... we are talking about the nature of this mega-economic, mega-ecological, mega-infrastructural control because the nature we deal with has this

character as well.

These articles, these books, and those working out the link between infrastructure/environment and our and other systems decisions is this DNA / DAA driver both of a social and purely environmental, material, technical, and economic nature.

It is also a direct occurring factor in the passage of natural scientific, and material processes.

This hyper DNA is present in every structure, process of nature, the micro-macro cosmos of matter and life.

It is this divine neo-evolutionary blessing.

The 10 percent of the neo-evolutionary share proposed from time to time is the seed planted in advance. Seeds of investment, .. .from / for hyper-evolutionary decisions to maintain, develop (for) maintain ... our position, our (literally) DAA bargaining card in an environment of potential and threat systems.

Not too short-term, temporary plans, i.e., a more neo-evolutionary approach to economic, social, ecological, productive, individual, family, national, global, cosmic undertakings.

We are not only observers but pre-all decision-makers (no decision is also a decision) responsible for ... saving ourselves in a crucible of cosmic vortices in which we participate more or less passively or actively ... it would be better to use the latter version of participation.

I am still trying to explain my position, role, and the position of my environment, system, and the world in the space of intersecting other systems and worlds with similar mechanisms of action and interaction.

Explanation of overlapping systems, worlds, and working out our position

to/re / neo infrastructure with the help of variable / mega DNA / DAA other systems in us and beyond directly or indirectly with more or less known effects .. ecological, hyper evolutionary hyper such as ... plane, rocket, not always completely aligned, with the same methods of influencing the worlds, e.g., systems that fix DNA (c) so-called maximum speed, DNA, the so-called life span...

We can, not only can but involuntarily take part in this inter-systemic and interworld process. The question of how consciously, actively, not / passively, and this still depends on our infrastructural, environmental, and evolutionary involvement.

Not only the question of commitment but also the application factor, the amount of use of the production tool more and more controlled, automatic, autonomous in its tendency ...

transferring production, mega-production, and neo-evolutionary competencies to a large extent simply to automatons due to the infinite enormity of evolutionary, ecological, and pre / genetic dependencies of matter and life.

This "10%" neo-evolutionary investment strategy consists of further searches, actions, and decisions of a mega-production, mega-ecological character, which aims to change nature, or rather its systems, as directly and effectively as possible, increasing the effectiveness of raising our evolutionary level.

These are not direct consumer actions but existential, over scientific, over economic activities aimed at more efficient material and tool performance aimed at changing the environmental infrastructure and further its pre / DNA parameters in the area of our interests, and existence, including classical science, economy ...

To be continued the drafts with the development of this 10% share in political and economic activities and using

this mechanism directly, by direct control of processes, systems, and structures of nature.

CHAPTER FIVE

Hyper naturally– neo evolutionary tool/model for the shaping/relating of any structures, processes.

The resurrections mean neo regeneration – EskmDAA> EvmDNA.

Elementary, non-relative/hyper relative model of investing/ processing/relativizing in hyper cosmic tool/material evolutionary systems – relations of mutual inter systemic materiality, objectivity as the basis, levels of generating the properties of the foundations of matter and life – in all initiatives, public and private projects.

Further analyzing, algorithmize, combing the ground, the area in shaping, relations, relations, and then shifting, transforming property boundaries in the participation of additional systems / systems as building blocks in the pre-generation of a new DNA model of matter and life, i.e. rearranging new laws of nature, i.e. overthrowing the existing system of natural relations to an absolute degree – not recognizing the current status quo of nature, the environment, i.e. direct competition in transformation, processing of relations between systemic object and subjective, depending on the role at a given level in further processing, neo / evolutionary structuring, neocosmic on physical and biological field, opening new foundations for the current practice of expansion, which is to overtake ... the expansion of the cosmos ... the expansion of other systems, including finally economic, strategic competition for establishing new laws, based on the fulfillment of the previously mentioned neo generation obligations in this mega space competition.

Yes. In the cosmic competition, i.e., creating new cosmic bases, material and tool bases as competing in the mutual crucible of matter systems and micro and macrocosm life – rocket, space base as a neo / evolutionary toy in the exercise, developing the base of its neo-evolutionary base to the above-standard material level (energy) ratios, EskmDAA> EvmDNA (Emc2).

There is a need for further infrastructural expansion, more and more effective in relation to the developing structures of other, often quite static, degenerative evolutionary systems inside and outside of us.

The need for further active (less and less passive) participation in the competition with other systems of the nature of evolution, which is a sort of squaring/modeling, closes our further development potential. They preserve and trap with a more extensive set of structural systems of potential.

There is a need to go (mental/economic) beyond these Emc2 / DNA boundaries to achieve supra-natural, supra-systemic goals with their disadvantages, disadvantages in relation to our existential, developmental (without development, there is no chance of existence in the dynamic environment of nature).

The need for a new mental, economic rocket of this 10% share in the search, in designing, interfering with the current currents of competitors, maybe not their elimination, but being on the same level or maybe with an advantage as a partner, a competitor in this, however, brutal competition between systems of nature, including the chain of surrendering existential addictions.

The need to break out of this slave mind, an evolutionary slave soul, for a soul that inspires a human being homo sapiens, a mental homo Erectus in relation to the plundered, subjected to currents, boundaries of other forces/systems.

There is a need for bold, systematic actions aimed at increasing our exports, and not only imports, i.e.,

interventions, investment changes in environmental change, contribution to its transformation and reporting, and not constantly importing, surrendering, subconsciously adjusting to its tendencies, i.e., not trying to a debtor, doomed to be literally grinded and eaten.

Is there any clear strategy? There is no. There is a thought, an idea, a concept, and a mixing in the environment of the natural systems. Construction as it was already in the continuation of the construction of new tools, and components for changes.

These tools are to be more and more replacing people, i.e., more and more industrialized, robotic, automatic, autonomous. They are to be our serving class of slaves, collaborators to total changes because only such have any sense in the total environment inside and outside us.

I know that many are struck... lazy, just thinking about the enormous systems of nature that we have had, we have to deal with. This immediately changes the direction of any thinking, including thinking in the scientific, economic, and civilization concepts.

Not everyone has to be directly involved in it ... but may have participation, awareness in supporting, costs of this undertaking, such as an anonymous voice, indirect or fiduciary (delegation of competencies... specialization, production, automatization, autonomy) participation in this undertaking.

In supporting programs, and projects, which are the most subject to competition, partnership with the systems of nature, that is, catching them up, outdoing them.

This partnership gives us the right. Practically partnership is the possibility of shaping the message literally, individually and globally, totally, materially, and in life.

The partnership is equal law, thinking in shaping nature, changes, and substitutes. Changes to its systems. We, of course, can also make an economic compromise on costs and profits of this mega-evolutionary venture, that is, not investments, research discovering nature, but changing it, or rather it's systems, or rather building our own, our own network of systems that will be able to grow, surpassing the existing systems. Arrangements that seem unchanging, impossible to change.

Neo-evolutionary automation, robotics, and robot cities are to be just the beginning of a new/different Big Bang, as if from the very thought – not even a point that has the infinite potential to change nature and the world.

Thinking in practice to build your own niche that can, without needing to, intervene everywhere in other systems, niches, and systems of nature.

It is this new relativism in creating, transmitting one's own thoughts, strategies, economic operation, and further technological and productive transformations in an ever more profound, more precise range of factors, DNA components of physical, biological, infinite bases, potential systems, natural components within us and beyond.

The need for organizational, political, and economic indicators of work, and costs to control this venture like the great structure of a neo-evolutionary rocket, neo-systemic in relation to the systems of nature in us and beyond.

Systematic participation in this general evolution of the micro and macro cosmos of matter and life.

It is about changes, shifting barriers, existential developmental perspectives, and finally, the absolutely infinite potential of the cosmos, which is contained at each point together and separately.

And here it is about technologies of exploitation of this infinite hidden potential not only in atoms, in DNA,

quarks, but also in their assemblies, their further internal and external inter structures and processes.

The issue of further production technologies, economics, politics, cooperation of thoughts, ideas, ideas for the further exploitation of the infinite potential of matter and life, literally and figuratively.

Ultimately, the aim of the articles is not some speech, sales, economic, scientific, and political career, but an indication of the mechanism, shaping tools, accelerating the transformation of broadly understood systems of evolution, attempts at transformation, strengthening the existential and developmental position in determining the properties, value of new models for triggering, indicating new potentials that are endless and only waiting for further technological, productive, economic, scientific, material exploitation ... entering this rhythm, a new rhythm, a new neo-evolutionary intelligence, relations, relationship inters, systems of nature, their existing models, limiters, closed circles, and above all the infinite potential inherent in external and internal systems as well as our new shares and contributions.

The point is that every enterprise will be able to find a place in this neo-evolutionary, supernatural model of existence and development.

These are the articles/chapters that search, pursue, that new generations, more and more new horticulture, forge, cast ... further components of matter from ever deeper infinite decks, potentials of nature, new niches, new structures caused by it, properties of models for the further expansion of a new one, already newer wearable / carrier, new DNA based simply on new perspectives, component parks, mega-production, which will continue to be the basis for further new models. Models more competitive in this inter-systemic competition of matter/life of the micro-macro cosmos ...

Yes, it is discovering, conquering, and creating a new America, new cosmos ...

That is, further beyond systemic, inter systemic, active multi-systemic discovering, working out, winning new social, economic, scientific, civilization, creation / neo-evolutionary space – hyperspatial evolution, and further production, thoughts, science, technology.

This above natural prospect shows as if from the outside the image of the absolute model, the axiom of processing mechanisms, the structuring of matter, which is merely a set of infinite assemblies of infinite systems of m, whose physical and biological usefulness of Esk depends on the depth of the DAA's own organizational contribution,

that is EskmDAA, from which m is the so-called natural background EvmDNA ...

Structurality determines the materiality, the properties of matter at a given level, which, if transformed, affects the next level of the structure, which determines the further stages of the new materiality, properties, modeling. That is, we are talking about variable relative ratios of any factors, depending on the structural or material level for further always changing structures on a given level and area.

This regularity can properly algorithmize, control, and undergo production processing, i.e.; we are talking about mutually responsible systems in influencing trends, tendencies, and properties for further structural levels, i.e., further levels of intervention in structures, processes, properties of matter, and life to infinity.

Each structure, old, new structure, complexes of old and young structures was, maybe, is the material for the next micro-macro level of the micro macrostructure, at a level appropriate to ruthless commitment. Participation in this process where a car powered by the same wind can become faster than that wind ... and any other structure and process submerged in the background with less efficient parameters.

Robots make robots to improve their parameters according to this over-natural algorithm of change/evolution

Systems make downstream systems with new properties.

It's about streamlining this process – permanent

instrumentation, mechanization, and increasing the totality and precision of this process.

Extracting this mechanism of tool improvement at every level and topic of activities, processes, and systems structures for their further efficient neo-evolution in the developed by these neo-systems platforms of further systems that often change the course of matters and models of tendencies.

Massive processing – it can potentially also, but it is possible to systematize, tame, and create new foundations for the levels of systems, which are to be the next element, literally a new element of the cosmos of matter and life.

Infinite sequences of potentials, trends of systems over systems, initiated by all processes, structures discovered, developed by systems, including our system

The question of finding this attachment. As they hooked, – using the DNA model, Emc2, etc ...

Manufacturing, economic, social, civilization, political, just macro microeconomic application ...

A set of such m systems can only be a component

to the basis of changes that generate changes, reevaluation of inter-system relations at a given level in a given degree and direction

Continuation of parallel worlds, systems, cosmos, which are a mutual material for their own better tools for their further neo-regeneration, possibly undergoing specific singularization again ... as fertilizer, material for re-use, but already on a different program, level, ... the new composition of bones, i.e., systems that may not suit us already in the given approach so far, or at least they do not suit me.

In the feast of the resurrection, neo-regeneration, overcoming all barriers of life and matter, indications of supernatural laws, pre-generic laws of matter and life, absolutely in relation to all physical and biological ties, I wish everyone the best in crossing nature according to the developed algorithm, the direction of structural changes jointly, or separate.

CHAPTER SIX

Supernatural activities shift the boundaries of our possibilities and chances negatively or positively in increasing or decreasing the widely understood relational coefficient of survival, development – an extended definition of carbon footprint and environmental composition.

Instrumentation determines the boundaries and shape of environmental structures/processes and increases or reduces the carbon footprint in not absolute terms but in terms of its benefit and position in relation to the garden of nature systems with which we maintain a specific economic relationship.

The benefits and profits of our existential and development position depend on the neo-evolutionary contribution, i.e., investments increasing our conditions and tools for changing parameters and levels of properties to our advantage. Non-neo-evolutionary, that is, consumptive, passive behavior in relation to the trends of the systems of nature in us and beyond, reduces our existential chances and increases the so-called carbon footprint to an absolute and relative degree.

Each so-called indissociable, elementary component of a given component of any EvmDNA model is infinitely subject as a material/potential (it is included as only m) further model divisions, laws, manipulation of EskmDAA due to external and internal discovered or generated process and structural states.

Each activity is preceded by imagination and mutually influences further processes and evolutionary structures. The greater, the more deliberate the transformation, productivity, the greater the awareness of the process of this neo-evolution, the greater the influence on the directions of the strength of evolutionary, structural, and process changes which these actions, imagination are, the pre-DNA, i.e., the pre-properties for a given environment inside/outside that is already different in which one stays in, influences, intends to change/move despite its often contradictory winds, trends, processes, and structures.

The so-called tricks, science, production, economics are absolutely only a mechanical component, a model of the environment, a given level of evolution – EvmDNA

Greater imagination, divine imagination, evolutionarily responsible imagination as a breakdown of its (EvmDNA) trends, its often primitive and brutal materiality/materiality/massiveness, increases the field of action, seeing beyond evolutionary illusions / reaching boundaries, or rather beyond the limits of any strings of matter and life.

Strategy, imagination, and faith define the limits, economy, science, economic, and scientific strategy in extracting endless layers of potential.

I have nothing against science, achievements that are a material part, inheritance, continuation of evolution. The issue of capturing, better and better grasping the model, the mechanism controlling this evolution at the production, economic and scientific level, where these achievements will be only (Ev) m (DNA) for further development and life -Esk (EVmDNA) DAA.

Here it is about a more comprehensive, holistic, or rather barrier-free treatment of the environment.

As infrastructural "m" to be further processed according to EskmDAA. No longer according to the closed model / closed models of EvmDNA, which are to be merely not even a material, but a mass for further free initiative ..., it

is able to exceed its ratio, determined by classical science. ... the law is there to be avoided ...

Treating the systems as material mass for newly excavated material does not always have to be carried out according to the existing systems in this material, mass, internal EvmDNA / Emc2 ratios ...

Holistic, that is, exercise, developing the perception of a larger collection of blocks ... domino Lego micro-macro of the cosmos – its evolution to take it over, overtake it ...

Not who on whom,

but economics bows here in the strategy of survival and development at the evolutionary or rather neo-evolutionary level... Only the advantage gives a non-degradative future.

Despite the previous relations, the so-called laws, their change is an expression of the divine, human, neo-evolutionary, economic mercy in favor of the law, the state ... according to the infinite potential of possibilities, the endless base ... beyond the law of nature – this is the EskmDAA.

It can literally break these pure consumer laws of EvmDNA – it is reinforcing with additional activities, EskmDAA initiatives.

Yes. Higher laws, that is, find, work out EskmDAA.

DAA is derived from Esk / m, where m includes EvmDNA.

It is resonating with ever-higher laws.

Creating conditions for this resonance to surpass them

Extending patterns over the current models' tendencies, adding their own extra macro-micro-macro environmental conditions like rockets in relation to wind and gravity...

The sets of overlapping sets, absorbing each other, the next is a continuation, with more and more holistic views, discovering the structure, shaping the structure, properties of this ... from our infrastructural/environmental "m" our effectiveness our exact DNA, but already pre/super DNA or DAA.

To be continued with the DNA/DAA resonance project.

CHAPTER SEVEN

Let us not look at nature as if from the outside, as an independent creation, system, but as a component of relations, i.e., economics, the economy of mutual influences in which we participate. Let us not try to adopt it – it is already not sufficient, unless as a partner to any changes that interest us.

Let us not forget that these partner/partners (environment/systems) and we-partners have an infinite potential for infrastructural and environmental combinations in achieving the changes that interest us to a virtually infinite and absolute degree. The issue of a combination-leverage system/domino effect in working out, and already more in producing, as efficiently as possible, appropriate advantages over the physical, material, biological natures, unknown natures of matter and life in us and beyond ... beyond, that is, factors that can contribute from inside to outside and vice versa.

Let the existing images and models not confuse us with the infinite potential of establishing the following models, relatively ... as beavers do in transforming their aquatic environment, and then in the subsequent consequences for them and their environment for us, positively or negatively.

Leverage, constitutionality, pre-generation of the domino effect in the progress of generating a chain reaction – just as other systems work mutually, we adequately relate in this section of the chain of relations. Find the suitable springboard, starting ramp ... not to find, not to discover, but to create max-min models and improve/implement them at a given level of this connection, the resonance of the reaction chain of the structure of matter.

Models of the so-called proper boundaries of the existing changes of the further transformation of the neo-evolutionary environment ...

Direct, productive, and further search in this direction, creating niches of changes with greater or less advancement, but certainly a necessary economic and scientific product of the future, the so-called niche ... of the microenvironment.

Creation/production of the environment, a niche of a specific environment in the system of ... our, a specific rocket / our spacecraft – a model that breaks through the current ... used items of matter and life.

Abstract models with more and more precise structures to break through the existing resistance existential models of matter and life – ..developmental always predictable – countability, the precision of creating one's own environment ... neo-evolutionary niche .... domino chain reaction evolutionary process

it has always been, is, and will deserve a niche; it will depend not on expectations but on the work put in, more or less active/ passive neo-evolution / hyper-ecological – production, protection, development of one's own niche of matter, and life, as any natural and supernatural system does it.

Generating, initiating, ahead of initiatives of other systems/ worlds, resonating in this way on the DNA of one's own and other environmental / world systems ... own development of these and that properties of DNA, and in this case DAA, to ever deeper DNA material layers in us, and outside the micro and macro cosmos to ever deeper decks. This is the only guarantee of not only development but often and always for survival at the individual and

collective level!, At the quantum, cellular, global, and interstellar levels, intergalactic, intercosmic, over the natural, supernatural ...

In every field of human activity and other more or less independent systems, more or less natural.

It is about the continuous increase/concentration of the EskmDAA infrastructure network interference inside and outside the structures, the so-called natural EvmDNA processes.

Productivity, efficiency, and automatism raise the level of this EskmDAA, i.e., hyper macro-micro economizes, learns, and processes of EvmDNA transformation of the current environment.

Where, how? – it is a matter of working out these pro-hyperevolutionary niches for a given direction, a given group of interested parties ... It is about a hyperevolutionary addition "hidden", "overcovered" in each system of the structure process ... The need to develop production conditions for discovering, directly producing, catching the given environmental factors such as the car to the wind, light and their DNA (which so far is faster than this wind), and literally DNA. Catching their wind, energy of a given potential in the sails, ... engines, surpassing it ... Such a superstructure of the DNA of wind, light, and other physical and biological processes ...

Just as the predominance of the human/homo sapiences species over the Neanderthal species was not based on the current physical advantage, but on the advantage in taking over, overtaking processes and structures of environmental systems.

This is what we are talking about here about the chances of superiority over the system, i.e., not trusting only the current levels of inter-system environmental consent but focusing on further evolutionary, over-evolutionary strategies in the systemic, inter-environmental, over-environmental, over-systemic, over natural race ... attacking nature ... building (EskmDAA) of our own nature on our own conditions, infinite potentials inherent in a-material / matrix / m, in us and the material/energy of this nature in every, every section and point, always.

Evolutionary superstructure, environmental superstructure, system superstructure ... These types of investments not only to survive but to increase the life potential of the matter environment by remaking it ... creating conditions for others to adapt to it ... of course by making it not in words, but brave over macro micro-economic ventures, micro-macro(macro scientific means neoevolutionary/over naturally) scientific ventures.

Do we have a program for this? Not. However, at least we already have issues, questions, or at least an awareness of the infinite potential of threats, as well as the infinite potential of opportunities for which it would be appropriate to have questions that are the keys to a good equation, and then solutions to an infinite number of unknowns, but also an infinite number of opportunities, where on a given level, we will generate them, generalize, exploit, transform them into a practical raising of the civilization level above the Neardental, above current homo sapiences, above the current level of the natural niche of the earth planet, the solar system, galaxy systems, quark systems, cell systems, etc.

The DAA tool controller will be this helper disengaging tool that will free us to raise, increase levels, potentials, ever deeper, dense the decks of matter and life.

These DAA, i.e., also artificial intelligence, i.e., technology subject to the infrastructural "m" in the EskmDAA model ... of holding a tool, controlling a tool/an enterprise, is building an advantage, superstructure, in competition with other systems ... over the Neardental, over ... homo sapiences and over other more or less advanced systems inside and outside of us ... often overlapping each other.

To be continued for systematics of pregenetic infrastructural recognition, control of processes-structures of matter and life in hyper macroeconomic and scientific activities.

CHAPTER EIGHT

EvmDNAn as EskmDAA as multilever/multitool of infrastructural/ environmental neo evolution of any properties of any segment, level, structure, process.

EvmDNAn as EskmDAA, as multilever/multitool of infrastructural/ environmental neo evolution of any properties of any segment, level, structure, process, model of matter and life by science and economy.

Increasing productivity is breaking down further barriers to existence and development. It is absolute neo-evolutionism, which is the basis of the leverage of the transformation of matter and life, even those who seem unattainable or from other worlds.

... The need to employ tools not only for the consumer but also for investment ones that further raise the consumer basis, the mechanism of consumption development – the consumer, scientific, economic base of the current observation, exploitation of the level of properties of processes, structures.

All in all, it is about the density of the structure of the given system with which we are dealing. Or rather, speaking more precisely about the construction of this system of matter and life, in the topics that interest us.

The structural directivity is not the mass, i.e., the so-called active masse (covering a given phenomenon) piercing its more penetrating structure by the process (energy) of classical gravitational and other loads.

Mining, manufacturing, developing the mechanism of an ever more efficient tool, and the DAA of this tool in conjunction with the "m" environment/infrastructure we have, in which this tool is further aimed at ever higher, deeper levels of neo-evolutionary intervention, which is a usual trend in the infinite so-called the natural network of inter-systemic relations that further magnifies the directions of the evolution of micro-macro cosmic processes inside us and beyond.

We function in overlapping systems/worlds, like inert or less, more influential air; in order to influence a given system or vice versa, it is necessary to pierce its structures, such as stone to wood, metal to stone.

Just like air, ... we are like passive air in the Emc2 system that is now practically, mentally locked in its known, perceptible structure and processes so that it finally breaks through it literally, chewing from the inside, gradually moving to unimaginable more penetrating systems that can this overhead structure breakthrough – this applies to physical, biological phenomena ... the issue of producing your own infrastructure, that is, your own environment to break the "m" barriers. Breakthrough as the systems of evolution/neo evolution do each other.

Everything is permeable. It is a matter of systematic imagination than the present infrastructural superstructures as it happens in the environmental development of our civilization and other systems (civilization, worlds) in the biological and physical process of evolution.

These harder infrastructures, more penetratingly developed, as if from their own neostructural initiative, can break through the old structures of the world, "air" loose structures, which so far at a given level direct / steer our thinking and action.

Productivity, efficiency, effectiveness, that is, eliminating new infrastructural tools is the proper process of EskmDAA> EvmDNA, which occurs in the infinite world of systems of the so-called more or less hidden nature and our system. The issue of determination of regularity and implementation in the schedule of neo-evolutionary initiatives allows for a real improvement of the civilization level of our life and matter.

It is supposed to be a lever of change. Literally, a lever, where the point of support for changes is the current environment/output/ infrastructure of EvmDNA / EskmDAA, material, scientific, and the processes around it, initiatives, systems of organization, control, automation, further instrumentation – are to be DAA, and then the effectiveness of this entire process presented on the attached outline diagram.

This tabula rasa is that "m" in EskmDAA, is that EvmDNA = m <EskmDAA. This "m", blank page, model, environment, the potential to fill for our activity, our commitment around this "m" employing the DAA driver, and Esk of this effectiveness, the economy of this process.

Everything can be further divided and reconstructed in the crucible of changing tools and structures, the infrastructure of a changing environment, in which we should deserve more and more / become aware of participating in this shaping of looser, as if impossible to catch, as if unattainable structures, environmental processes (usually "masses"?)

The limits of the environment's possibilities are currently achieved after the transformation of Emc2, which is this border pattern. Still, we can further add over-patterns, sub-patterns of new structures, and older world structures within and beyond the scope of a given environment/infrastructure to these existing EvmDNA patterns.

We are between the systems in relation to the downstream systems of the micro-macro cosmos.

We will expand our significance and position by building and developing neo-evolutionary and infrastructural levers for EvmDNA <EskmDAA, where m comes from EskmDAA, which is only or rather the entire

structure/process of EvmDNA, which is this potential, a barrier to gain.

Neo-evolutionary activities are superior to macro-micro economic and micro-macro-cosmic activities at the scientific, economic, and ecological levels.

They establish the passive and active directions of the source of the level of development or structural and environmental degradation.

These neo-evolutionary economics, science, and principles are at the gate of mechanical evolutionary development, economic matter, and life.

According to these rules, guessing, finding, and production of all structures and processes occur.

Creation of an algorithmic machine, which is to contribute to a more free juggling with structures, processes of producing a specific resonance domino effect in more and more further, more profound, denser structures/systems of the space of matter and life, which are to influence more full responsibility and, at the same time, the power of commitment, the forces of influence.

They are not to be modeled, but they are the basis ...

the question of sailing between forces and systems of the micro-macro cosmos ...

To be continued in construction in this neo-evolutionary war of the ocean of worlds, systems, ...

To get out of this complex perception, technological, physical, biological, a specific gravitational bubble that occurs due to our civilizational back

To get out of this complex perception, technological, physical, biological, a specific gravitational bubble that occurs due to our civilizational, profound technological, mental, and technological backwardness.

We continue to comb the material further, create a new one, and grow our own soil of matter and life.

For the further, deeper infrastructure of our own DNA, pre / DNA that is DAA administration/management of neo/evolution of matter and life to an absolute degree to get out of this complex perception, technological, physical, biological bubble, a specific gravitational bubble that occurs due to our civilizational, profound technological, mental, mental and technological backwardness ... as it was with the observation of the solar system before and after

Copernicus.

It is about technology and production precision as a tool in determining and flowing through environmental structures, its local and global infrastructure on the more and more macro-cosmic, more and more micro-cosmic level.

We are talking about non-relative mathematically, hyper/evolutionary models, which nevertheless determine the usefulness and directionality of the process of applying other models.

One can attempt a pyramidal classification of models, overlapping groups, absorption of one by the other due not to error, but a further level of penetration of the environment, the level of interference, manipulation of the

use of technological, production, and strategic leverage, i.e., changing the environment, courage in this process of changing, initiation of transformation = neoevolution of the environment in us and beyond to the extent of geometric progress – that is, in a few years we should decide for ourselves about the evolution of not only the technologically known planet Earth, not only atoms, not only cells but the entire cosmos known to us, i.e., in fact even more than the micro-macro cosmos we know today. Potentially, we can already begin this process of level 10 civilization, which changes the course of thinking, existential and developmental action, no longer ordinary homo sapience, but a homo creator on a scale surpassing the boldest futuristic and science fiction imaginings. We can potentially begin this process of level 10 civilization; it changes the course of thinking, existential and developmental action, no longer ordinary homo sapience, but a homo creator on a scale surpassing the boldest futuristic and science fiction imaginings.

Will we take over the functions of Big Bang/nature/fate? Yes, of course, we will not constantly watch and justify ourselves with it.

The breaking of the monopoly of models gives, opens the known gates of mass matter, new potentials of matter, material, and so further, life; as if you were controlling an ordinary toy, a drone, but already on entirely different levels, the massiveness of application, and visualization of this process, as it was and is just that at a faster and faster and more autonomous level, which will give freedom of development potential as it has given and gives the mechanism of a free-market economy of nature and beyond.

We subject them to a critical, inspiring assessment and then use them from things that are. Thanks to this, we produce items further, i.e., actions, not theory and DAA's imagination/initiative, which is more important than the knowledge/achievements of the current DNA.

The more tests, structures, and models, the greater the chance of completing the relogation, changing the DNA of a given area of the acquired DNA, i.e., pre-DNA – it does not even go about changing this DNA, but changes in the internal and external parameters of the structures affecting this DNA, which will allow this broadly understood DNA to play its role more effectively – i.e., external, internal infrastructural, material, specific gravitational factors pre DNA, i.e., infrastructural superstructure, foundation DAA, i.e., EskmDAA

it approximates picking up partial vector redirection of tendencies, the evolution of matter and life

to an absolute degree – relations, relativities have nothing to do ...! Locally, there will often be other effects, the reception of resonance or non-resonance of the action.

Good empowerment, economization of relations to a great extent, relativity neutral

(economics of the advantage of exchange advantage) noticed with the appropriate use of instrumental infrastructural/ environmental compaction, which influences the potential of the broadly understood mass (gravity) of environmental matter consisting of other systems less dependent, and less passive on our influence and action.

By influencing, processing, and animating, we increase our environmental contribution and thus the potential of the elements of matter and life that interest us.

For this, we need more and more efficient tools and organizations so that they do not become passive waiting, shifting tasks and blame on nature or god – but work and its effects of our actions of a degradative, evolutionary or creative nature – neo-evolutionary – neo-environmental, neo-natural, which completely has to change the course of events in space and matter and life in terms of processes, phenomena, processing on our part in a really micro and macrocosmic perspective, in the shortest possible time.

The mechanism of increasing the efficiency of the EskmDAA process affects

the gigantic dimensions of the challenges of the venture ... in a few years, the mastery of the universe so far known ... But we must get to know and, above all, go beyond its models EvmDNA = m1 = Esk / DAA <EskmDAA1 ... n to master it in the micro and macro, cosmic, biological, physical, in the dimension already over m.

It is moving fast to higher, deeper, more profound, and further dimensions of precision, the efficiency of the infrastructure base, its constant expansion in a more and more tool, production, automatic, autonomous way, liberating our minds to even larger, more and more distracted from mental, mental gravity EvmDNA.

It is a neostructural, neo-environmental law/process of influence/initiation / neoregeneration, i.e., the pre-generation of the structures of matter and life processes.

Each neo-structural change contributes through the densified structural interference on the DNA of our, as if passive from others, the "air overhead" system of matter and life in our world, the cosmos, which may further contribute to the productive use of this phenomenon to overcome structurally denser/more gravitational systems which generate the physical and biological properties of our current world environment.

Just as the free market revives the economy, the release of the neoproductive, neo-infrastructure, and neo-evolutionary system revives the grid, compacts it as more potent, a more substantial base in the dense, loose "air" space, and its more dense objects, systems, pillars ...

Our mathematical approach to these phenomena breaks the horizon of events; in this case, a deeper application of this mathematics in models, and formulas, further outline the infrastructural model of the environment according to EskmDAA and, at the same time, its practical application.

Winning or losing in this process is further steps backward or forwards ... every activity, recoiling push in the infinite potential of the environment.

Continuation of the systematics of the pillars of the neo-creation of matter and life here and now, immediately, not in the distant future

Mathematical – any so-called model-based approach to phenomena helps at a given level to understand, apply a given process, structure, and at the same time it can create a certain tangible optical phenomenon, illusions that everything in a given topic has been counted, and after all, each phenomenon has a further infinite decks, levels of natural, technological potential, which cannot be determined by one constant and infallible pattern at the same time, unless with unknown data always entered into further metatmatic equations – here it is all about breaking down barriers seemingly natural and technologically insurmountable due to the direct illusions of the scale of a given phenomenon, or rather an illusion, the attitude of the lack of its scale ... an internal, external barrier affecting a systematic, successive approach to this new phenomenon, process, structure, exceeding the existing biological, material, microcosmic, macrocosmic evolutionary phenomena – that is, we are already talking about mathematics going beyond its parentheses, neo-evolutionary mathematics, where the previous EvmDNA of matter and life is contained in m, in the neo-evolutionary, hyper-natural, hyper-technological, hypere-economic formula EskmDAA, and this is further contained in m in the formula EskmDAA2 and so on, which changes the mental static / mathematical approach to the phenomena of matter and life once and for all ...

CHAPTER ELEVEN

Hyper mathematical, hypernatural, hyper-technological, hyper-economic model of neoevolution of life and matter for changing...

Hyper mathematical, hypernatural, hyper-technological, hyper-economic model of neoevolution of life and matter for changing, breeding, producing, creating, developing, processing the nature of the cosmos in an efficient and immediate degree – m1 as EvmDNA nature, environment in the formula Eskm1DAA (+ -) n.

Infinite and overlapping contributions to the behavior of matter and life, their structures and processes, and thus our and other contributions, as if from nowhere. The development of this contribution as precise and efficient as possible in a multitude of other forces and factors.

Always each pattern is included in the EvmDNA model, which is m1 for the EskmDAA body of overlapping natural and artificial systems, discovered and undiscovered, invented and not invented.

Construction of the mathematical model EskmDAA (+ -) n sail of neoevolution, neo nature – making/developing our own nature and its properties to strengthen our natural "nature" / system – which is weak/small in relation to further and closer micro cosmic macros dimensions of threats and competition – we are to be the initiators of changes on a scale exceeding any scale.

A mathematical cover as a shamanic cover in a variable grid of the level of dependence, interference, superstructure, foundation ... the same pattern, e.g., a physical, mathematical representation of a given process of physical structure ...

It may or may not be significant for the data of the same other physical, biological occurrences, including the interpretation of the physics of time.

Additional "artificial", "natural" factors may distort the model of any previously established laws ...

Overlapping physical and biological models, and hence feedback mathematical models, which often contradict each other depending on the level of interference.

Further technology – i.e. manual shifting- influences this operation process through and for the benefit of automation, i.e. the neo-naturalness of processes and structures.

Each model and formula factor has its primary EvmDNA and secondary EskmDAA ...

Environment, surroundings / EvmDNA define us; our being, is our reflection.

As we treat it from the inside, from the outside, we will strengthen or weaken its properties, our properties.

We are also talking about the factors of the so-called external environments that may favor or interfere with our development environment ... including technological, economic, social, and political factors.

That is, literally primitive protection of the environmental status of ecology is not enough; it should be constantly strengthened and to a maximum extent to achieve even the smallest progress of protection and development, development to protect us, our local natural system, hyper immunologic, surrounded by a gigantic dynamic ocean of competing environments, systems, worlds overlapping each other and our world of matter and life – we must mentally, technologically, economically and scientifically come out of the vision of the world we see here and now ... because it is too little from the point of view of individuals and the civilization community.

How little we feel, how little we act, so weak will our environment be, and vice versa ...

How we support the environment – excessively, this will be the guarantee, the foundations of the environment of matter and life. We are talking only about the over-barrier approach because only such a vision and action guarantees the recovery of matter and life in our system of protection and development if it is to be effectively continued at all.

The environment, its change affects the variability, changes in its properties and the potential that we are, we can be an active or passive shareholder, but only on a large scale of vision, decisions, processing, structuring ... because we will close ourselves in the shamanic square Emc2 literally and figuratively ... these visions are coming out, breaking down barriers above the technological bubble of the perception of nature ...

We have to manually and consciously control the invisible yet visible hand of the environmental market, not counting that it will somehow sort itself out ...

will fit into the mind of the competing systems of nature, ... just like between an antelope and a lion ...

It may be a compromise, but it will not happen by itself for our benefit, but the benefit of the asteroid of doom, the road roller of the evolutionary cosmos of matter and life.

DAADNA steering supporting economic activities, i.e., processing of neo-evolution and at the same time properties .... DNA <DAA of matter and life, new matter and life as a new dynamic building material of matter and life, the basis of properties other than light, because different in deeper layers, layers, causes of

matter and life, today interpreted according to a closed, degradative technological/environmental bubble, and thus a retrograde mental bubble ...

To be continued in unconditional, not so relativistic/dependent economic, scientific active contributors/core management, processing, neo structuring with the transformation processes of micro macrocosms in us and outside.

CHAPTER TWELVE

Natural or caused by our ingestion, the asteroid of collapse and destruction EvmDNA or the asteroid of protection and development EskmDAA.

Natural or caused by our ingestion, the asteroid of collapse and destruction EvmDNA or the asteroid of protection and development EsKmDAA. We must be more powerful than nature to survive to live ... EskmDAA> m = EvmDNA.

This asteroid is either Ev-economics / efficiency of neutral subordination or Esk – an initiative that strengthens us by strengthening our own environment/infrastructure of matter and life.

Esk – Selective economics, always superior to primitive Ev – economics and science of environmental neutrality = submission = indifference and makeshift. But Esk is becoming very fast like Ev, so it requires powerful and fast, ever more efficient actions and reactions to keep it from becoming (-n) Esk Ev.

Consumer technological gadgets passively and reciprocally drive the mainstream of hyper infrastructural evolution of matter and life. This current is a passive mass and, at the same time, the potential of neural connections in macro-and micro-cosmic terms.

The issue of proper constant monitoring, control, tooling, economization, technologization of productivity, automaticity of this process, which is passive and active depending on the level of intervention – not only artificial but natural ... we take part in it consciously and productively. At the same time, naturally, nature in us and beyond needs us as a salutary contribution to the gardener ...

Gardeners who pretend to be innocent ecological worms do not guarantee the development and safety of the nature garden against "asteroid" dangers. It does not give a chance for an "asteroid" neo / evolutionary breakthrough, going beyond the brackets of technological, natural, mental bubbles, quadratures, "Emc2 – closed squares ...

This suggested 10% core of all entrepreneurs participating in the human anonymous, influencing the flow of potential, economic, economic, hyperevolutionary environment is to be that conscious, neural core, action, direction, direct management by each individual, non-blind dance, process control, and interest directionality in controlling neo-evolutionary factors in economic and scientific activity, which is to be the key to go out from the mental, technological and perceptual bubble of security and development in the ever-faster transformation of matter and life at an ever further, higher level ... where the supply will exceed the demand of our dreams. ..

We have to kill all the jungles of the cosmic system, kill the cosmos in order to surpass its properties of matter and life forever ... maybe not so much in the first part of the sentence, but definitely in the second part ... and we'll do it for sure ... soon ...now....

We have to go beyond the square of the parenthesis, the closed circle of disabilities. Create models that go beyond closed medieval visions of some cosmic impossibility.

We have to accept that not only we can, but we must go beyond the boundaries of this system, the world, and the micro-macro of the cosmos; it seems that there could be technological, leverage possibilities that could not potentially affect the changes exceeding the scale of our imaginations, perception ... and the scale is truly infinite, infinite structures, processes of matter and life.

And we are to adapt to such a hyper-scale because the scale of the potential and threats of the surrounding life and matter is also infinite. Whether we like it or not, we are passive, more or less conscious participants of this endless process of neo/evolution.

We must take this into account in our projects, patterns, models, thoughts, imaginations, aspirations, and fears, as the core, hyper DNA processing of environments, systems, worlds, cosmos – in a civilization, individual and concrete perspective.

Inserting, fixing, processing, working out, extracting, and saving the dynamics of Eskm1DAA in each previous pattern hidden/ closed in m1 as EvmDNA, as Emc2, and the like.

Fighting for oneself in the so-called natural system, where every over natural initiative is a further extension of the self extra in nature in its transformation process consisting of these initiatives. The question of the mechanism, improving the tooling of the interests of this process on our part, its economization, production, massiveness and automation and support, incentives for the best initiatives, own artificial natural selection, own evolution in an ever faster and more spatially powerful pace, in an equally fast and powerful transformation environment other competing systems, also potentially cooperating with our economic, scientific background, system ... and we with them.

How to get out of the EvmDNA bubble of technological, mental, economic, and scientific level in terms of EskmDAA, where EvmDNA is only m in EskmDAA – new efficiency, Esk leverage, and new decision-making DAA multiply, surpasses the current effects, Ev achievements, facilities, infrastructure of the old "m" and previousdecisions and attitudes natural and artificial DNA ...

CHAPTER THIRTEEN

The spectrum of the infinite perspective of undiscovered or artificial elements of EskmDAA.

The spectrum of the infinite perspective of undiscovered or artificial elements of EskmDAA and the existing carriers of elements and laws as divine elements of creation, neo-evolution of matter, and life. EskmDAA (-n) <EvmDNA (Emc2) <EskmDAA (+ n).

Evoking spirits of matter and life ..., removing new elements from the ore, and further in this process through active infra structurization – depletion. Discovering and inventing new artificial and so-called natural. Simply producing new foundations, new elements, new bricks, new elements, new components, new decks, giving further/broader foundations of the laws and properties of matter and life, is to be this initiative, imagination, and idea by no means surpassing the current, literally environmental, infrastructural level, and based on it his knowledge.

DAA or divine assuming the helm of responsibility as a neo-evolutionary process, imposing new trends in specific areas of evolutionary transformations of matter and life.

DAA explains the so-called divine contributions of transformation, and creation, that God, i.e., the personification of this process on systems, and people, which influence further changes in lower systems

Imagination, i.e., DAA higher than the existing DNA of matter and life, i.e., this divine element of the

undertaking, initiating changes as if from nowhere, although they are always dictated by known and less known contributions, systems, ... worlds.

This divine, supra-systemic, multi-layered, multi-contribution approach to processing and structuring matter and life completely and totally breaks the principles of the so-called chaos system, a system of primitive views of the environment in the so-called absolute models, physical and mathematical patterns, and at the same time opens up infinite possibilities of using the right to it directly affects all the processes and, further, the properties in terms of micro and macrocosmic, in terms of more economical, political, mental than physical, mathematical chaotic or exemplary, statically religiously passive.

So it is not religion, science, or economy in terms of a passive view of the environment, but in the sense that we are gods, children of God, and we are fully responsible, and at the same time, the potential for a total transformation of the environment, infrastructure, environment, infrastructure on a natural scale, or possibly as massive as possible.

This is heading toward breaking the monopoly of gravity-resistance DNA background for matter and life on a mass scale. Mega economy and mega computer science are in overdrive, managing the evolution of all along the way of the micro-macro space of the cosmos.

So this is just a continuation of unconditionally putting directions, DAA signposts in the reproduction of the cosmos in us and outside.

This religious, over-natural, over-physical imagination is more important than knowledge; the so-called hidden, anonymous (without direct physical, mathematical, economic factors, causes ...), a pregenetic initiator of further decisions and process direction.

Knowledge is the existing DNA of the information resource of a given system's matter and life processes, and the imagination is the following, contributing external and internal factors resulting from further modification, such as the contribution of our DNA and other systems along the way.

Creativity, imagination, and behind it specific DAA pre-initiatives influencing, controlling, and determining DNA is a salvation, a chance to come out of the systemic, organic, gravitational slack ... aging, wear, and tear in a dynamic infinite net of gravitational inhibitions, canceling each other out, inhibiting each other, or accelerating natural and artificial systems.

Creative freedom, innovative, neo-evolutionary, neo-environmental momentum, i.e., neo-infrastructural, neo-economic, neo-productive, neo-scientific, and support for total robotization, automation, autonomous, neo-generation of the environment in a possible entire autonomous mechanism, neo-natural ... starting, imposing, your own environment own pre generic DNA of the development of the micro-macro cosmos, it is the only chance and basis to deserving survival, individual development, civilization against the background of other microsystems of the macro cosmos of matter and life in us and outside.

The point here is not, in fact, an anti-ecological attitude towards our environment, but on the contrary, its hyper-active reinforcement against biological, material threats of an apocalyptic, degenerative scale, which emanate from genetic and pregenetic systems of matter and life overlapping, mutually pushing out, competing with each other, against each other.

Liberation from slave adaptation to cooperation ... they, other systems for their world, and we for our world we will support the models of development that interest us, always a step forward, at least from our point of view of interests.

To be continued with the approximation of the spectrum of the mechanism of formation, discovery of old and neo elements of matter and life in an infinite scale of transformation of the background of the environment and its, our infrastructure.

CHAPTER FOURTEEN

The spectrum of the endless system of process/structural flows of matter/energy (and life).

The spectrum of the endless system of process/structural flows of matter/energy (and life), as an image of any additional processing m natural and production model in terms of = Esk/DAA.

It is in everyone; it is in each of us; there is bursting hidden energy in us. This is a force more significant than the strength of black holes. We have infinite potential. The question of the right line, direction, flow to which we shape, we produce tools for further ... production of the exploitation of endless energy of matter (life) organizational, economical, infra-structural environment of the micro-macro cosmos.

Infrastructure process flow systems.

Structural, which is enough to manage properly, tamper, accelerate, automate, produce neo-infrastructure, environmental, neo-environmental

The spectrum of structural flow and the mutual factors influencing its effectiveness at a given level .. according to DNA/DAA spectrum of the structure of matter (mass) m according to DNA/DAA, according to its properties, effectiveness.

According to DNA or DAA, the properties of Ev/Esk and infrastructural flow are about a system of decision-making, efficiency, automatic/intuitive awareness, and decision-making neo-autonomy.

Further searches, suggestions for additional shots, the graphical object, model, pattern, word, a sentence for

clarification, processing, pointing to the direct effect of the initiative to shape the environment, the actions of which have known and unknown factors, DAA impulses for further changes in the environment/background, its properties/efficiency, its economy, that is, its different Ev or Esk. And it is simply one's own decision based on the current level of exploited, organized, interpreted, produced, artificial, natural, artificial potential, more or less according to established laws and models. Simply a decision of the DAA resulting from effectiveness, from a natural, infrastructural resource, including a scientific, economic, civilization, and real-life decision. The decision, think on the dynamic spectrum line of the processing properties of the infinite potential of the micro-macro cosmos.

We must create open models for further more effective, non-passive participation in the open model of neo-evolution and not be degraded in order to maintain some value ...

This is to give an advantage over the constantly collapsing modern EvmDNA system ... space-time DNA – for mass rebuilding, or relatively parallel construction, modeling of the force system ... extending the spectrum of activities through the constant reconstruction of the environment, its natural and "artificial" infrastructure.

What does it mean? The fact that own projects, initiatives responsible for further changes, i.e., active responsibility, is a process of creation is divinity, which is the decision-maker in processing the directions of the development of the environment of matter (and therefore of life).

Any actions, influencing processes have a variable nature of environmental processing; they are divine and ruthless (but also merciful by the strength of the decision/power above law, changing the systems, models, patterns of nature's matter, and further their properties/effectiveness Ev <Esk resulting from the variability of decisions DNA <DAA in natural or artificial processing of m infrastructure the environment.

That is, dynamic administrative activation of the DAA of control as a pre-DNA, which has the past character of its "CV" and a future, futuristic, active one.

Divinity means greater awareness, inertia in relation to anything else, including physical laws, models, and exemplary (always going beyond the brackets of a formula, m model, i.e., according to EskmDAA ... phenomena and structures, with a variable external DNA image of phenomena ...

Reversal of the method of acquiring, transmitting electricity, transmitting matter/life, and its processing.

Activities, initiatives, the area they occupy, space, scope, and their limited spectrum is determined by temporary physical laws, specific physical properties-effectiveness, and other data structures and processes.

We must also take into account the so-called acts of DAA. Let us consider this DAA because DAA is GOD; it is divine, from "nowhere", from further deep factorial regions to the nature of the environment, our infrastructure, and other systems that should be listened to. This is the active faith, activity, and listening to the sounds of the needs, opportunities, potentials, and threats – We can be not only passive but active, the awareness of external systems and their systematics.

Our initiative and the initiative of other systems.

And this is the essence of creation, creation, responsibility for everything, because everything influences us, is next to us.

To be continued with this rebellious DAA> DNA approach to nature/infrastructure m – not demanding, not thankful, but co-responsible for its property, efficiency Ev / Esk.

CHAPTER FIFTEEN

DAA is the creation, further continuation of your own code ... like the evolution/creation of other systems' codes, and DNA is a given existing code, which is expressed in an infinite spectrum of wavelengths, reflecting, reflecting a given process structure, the rest of this pattern ... EskmDAA, EVmDNA. We are merely an invisible point. Our world is also an almost invisible line, a color in an infinite palette reflecting the structures of systems, and worlds, expressed in wavelengths as well as other factors, indicators that we can use more and more extensively and actively in our aspirations, possibly less ignorant, less arrogant to this background matter (and life).

Wave spectrum as a code in discovery – a reconstruction of the environment into a closer and further distance, depth, size, precision, and other more or less significant physical, biological, and technological factors.

Animating this spectrum, stimulating it internally, expanding it with your own technological, environmental, infrastructural, economic, ... and feedback, civilization, political initiatives

There is a need for our own contribution to more or less chaotic factors in the environment for artificial and natural insertion of our own vector of transformations, evolution, and creation

as a natural creator, divine over the following systems within and outside.

This own factor has to be extensive and accurate. Customs because it is developed as a real decision-maker of relations, which in the spectrum of infinite lengths, as internal relations along the length of these systems/worlds, where we are only a micro-line in this spectrum (DAAn) in length and depth exceeding minus and plus its dimensions to infinity.

Invoking, introducing into the code/spectrum of DNA / DAA the nature of other systems, own code, building blocks, changing the so-called constant physical laws and other dependencies.

Introducing changes as if from so many through theoretical initiatives from nowhere, from an anonymous environment ... our initiatives for the environment, not having to have anything to do with it, of course with always more profound changes also based on unknown factors contributing to initiatives, action from "nowhere" – of course, based on infinite cause-and-effect relations as a result of which and for them.

DAA creationism and DNA neo-evolutionism is the result

Esk of selective and creative economics of direct decisions resulting from "m" factors of the past and planned future.

Ev passive evolutionary economics resulting from a given infrastructural/environmental system ...

Evolutionary economics of matter (and life) Ev, a

Esk creation economics – is about increasing infrastructural power in an evolutionary-creation relationship, not in passive hyper-evolutionary ignorance.

We enter our scope of operation, correspondence with the economy of matter (life), or rather we cut charts, interests of other systems, and the so-called external physical and other laws.

We enter the field of creation because this is how systems of matter and life arise, resulting from the interests of more or less intelligent systems of the micro macrocosm.

Earth is the cradle of humanity. However, you cannot live in a cradle forever. The same applies to the matter systems and life of the micro-macro cosmos, its laws, and relations, which can be freely changed indefinitely. This is dependent on the economy of processing, transformation, production, massive change, i.e., more and more efficient tooling to take over responsibility, and thus the law, new laws (property) for ever wider spaces of the environment "m" and its properties/effectiveness Ev < Esk, and not only thanksgiving ignorance of the existing and yet dynamic physical, biological, chemical environment ...

Technology, the economy of environmental technology, is the determinant of all changes that have taken place, are taking place, and will take place so far, and thus the interpretations and usefulness of using new models, the law of matter, including life (although life also determines in return further neo-evolution of the structure), property/efficiency of matter m according to EvmDNA = m for EskmDAA .... the so-called materialization of decisions) ... and then economics, which again affects the productivity, equipment of science and its further impact on the economic development of civilization ...

Creation and evolution ... what was the first egg or hen.

Of course, we are talking here about our co-responsible influence on the environment, that we are this hen and not only an egg in the processes of transformation of creation and evolution, where the so-called material laws and life interests of the so-called gravity systems of the micro and macro life of the cosmos intersect and clash.

The properties/performance of the egg to be influenced by this hen can be shaped separately. These efficiency properties are nothing more than the economics of controlling processes, infrastructure structures, environmental infrastructure, equipment, maintenance, and development of this process in the right, non-degradative, passive direction.

To be continued for the performance of the neoevolutionary economy share for/by everybody and his

animous/creative decision...

CHAPTER SIXTEEN

Passive reception of impulses, signals from the micro-cosmic environment of worlds/systems, but an active reaction, not only in reception but in transmitting (literally and figuratively), imposing your own wave field of your own infrastructure, thus affecting the properties/efficiency of the environmental infrastructure, systems, worlds, which also include, among others, our interests, our matter, our life, our infrastructure, our ecology, our economy, our physics, our biology …

That is, direct or indirect infrastructural participation and extra participation in neo-evolutionary processes. Always DAA over DNA, it is this divinity of creativity, and divinity is a responsibility because, without obligation, one has no right to … exist; there is no way of defining, influencing the laws of physics … according to our desires.

Of course, we are talking about active infrastructural responsibility. The larger the infrastructure, the greater the actual responsibility, the potential of responsibility to control processes, structures, and further properties, and further economic, technological, ecological, physical, and biological effectiveness.

Responsibility hyper ecological, hyper environmental/ infrastructural but

Only a very active, robust economy, and then science and economics give the foundations for development; if it, this responsibility will have the features of conscious participation, and not of an empty social "market" game, which gives potential, but primarily idle.

Maybe it is, or even necessary, a kind of play with a zero, indifferent/neutral score, which means automatically a degradative, in a powerful, dynamic mill of rolling lines, grids

dependencies of micro-macro systems of the cosmos.

Matrices of the so-called cosmic waves of the DNA environment should be imposed on the matrix of our actions – not always overlapping ideally. Still, these programs should be directly or utilizing re-infrastructure restructuring of our environment economically, mentally, and politically, such as infrastructure, environment equipment ... like beavers ...

It is about taking the helm of evolution at the involuntary highest level, that is, responsiveness, which gives us the right and duty to protect each other's environment of our life, existence, and development – to maintain by developing.

This role must be felt by every citizen, client, and commander, directly giving him a tool for anonymous depriving him of specific support of obligatory, joint-stock entrepreneurs selected internally, not only evolutionarily, but also strengthening the economy, science, giving them additional, more conscious alternatives to real development-existence, so that the cancer of unauthorized actions does not have to appear to us like this, threatening ...

So faith in god nature is simply an active responsibility, not ignorant passivity, ignorant self-determination.

Expressing itself through courageous undertakings, literally and figuratively moving the earth, the cosmos, the micro-macro cosmos of matter, to further create the world from this clay/matter/mass in the image of a god ... responsible.

There is always resistance to making any changes or moves. Still, luckily, never our own (and other systems') initiatives, movements are never completely neutralized against them in further evolution, neo-evolution, and creation.

Despite these so-called mutual resistances, we always come to some changes. We do not stand still, even when we do nothing, because other participants, more or less material or alive, constantly influence further changes – the direction issue, so the changes do not have to be negative consequences.

The more a person takes an active part in this further exploitation and infrastructural transformation of the environment, the more he has the chance to find his own path, an initiative in this universal infinite set of systems of transformation of matter and then its DNA and then rather earlier through our interventions in this DNA, becomes a component of our programming, infrastructure initiative at a mono-factor / product level,

or simply as a kind of external addition, the sum directly raises the level of artificially platforming skeleton structures to the above-existing state.

So our intentions supported by actions are a direct or indirect component in the neo/evolution of the creation of environments. There is its basis – the additional pre-DNA, complementary DAA, in some cases replacing at a given level of intervention.

They are a creative soul supported by responsibility supported by deeds, which gives the values of the law and laws of existential development – the responsibility of development/action.

To be continued with the fact that every pattern, a model can be an active subject, is always somehow an active subject to manipulation with or without our participation in the infinite mesh of intersecting systems, infrastructures of matter, and life.

CHAPTER SEVENTEEN

by spectrum for the built spectrum

So modeling ... hit at random in a given topic and direction and, according to this, the exploration of this DNA / DAA construction into the following spectrum, which may be helpful as further, higher drivers for one's own and/in this neo/evolution environment ...

This systematic modeling increases the chances of effective recognition, creating our own spectrum as a determinant of further production, automatization, economization of digitization, and systematization of DNA tooling> DAA – generating our own bypass processing of the nature of the neo-evolution of matter/life creation.

It was so creating things that were not possible so far. – always impossible this way, but possibly another way. Let's expand this road network of infrastructure/neo-environment to make the impossible impossible. We can change the world/universe/ environment of life and matter by its constantly rebuilding road network of structures and processes of transformation ...

We live in a world of intersecting systems where any material, biological, physical, productive, tool, an economic, or ecological state may have a different value than the one we calculate that we will work out according to the chosen, existing, so far proven linear calculation.

And each time, we deal with factors that, if not thematically, may affect the state's result.

We must, privately, publicly, economically, and scientifically, always have this network of intersecting opportunities and interests of various systems, worlds, and sequences of conversions in our eyes. For my part, I always have this goal, even the most unrealistic one, which, in this infinite network of direct and indirect connections, always gives me a chance to achieve it.

One must try to traverse this grid, add to it, refine it, work out, and enforce it according to a possible confluence of these intersections of known and expected lines, new calculations, studies, opportunities, and threats by establishing and executing, one's own path through often not, and mostly unknown, intersections, connections of phenomena, processes, structures, events. This mesh is to go beyond the hitherto linear natural and artificial trends.

Coming out of our own life streak, literally and figuratively, is to use these levels of this reticulated path of development because, without development, we cannot even maintain a given status in a dynamic environment. And we mean not only primitive agonistic vegetation but development and creation.

All successive economic and scientific undertakings have always been based on going out of the existing systems and creating our own network, paths along with the network of dependencies, and possibilities for making our own net niches for development and invention, where the same things took on a different dimension, potential in a new perspective, network, new path execution of this shot of a new, new structure, new process, new production, new venture, ... more supported by a broader foundation, mesh, niche.

We are talking about more or less systematic or so-called random factors and introducing an additional contribution to support appropriate changes to a given point of intersection or connection.

Like a cancerous hypertrophic cell in any system, the model must and is subject to influences or machining. It will not always be sufficient to sustain itself in a changing reality.

Need for flexibility, forced flexibility for survival and development ...

an additional always newer factor to an already ... new and already old model ...

Each pattern, a model, can be subject to any manipulation ... and it is constantly subject to manipulation ...

Each process and endeavor may be different from the first but just more primitive than the next ...

Economics for physics and other fields and feedback

as well as in the next ones, it is already primitive for the following.

every field of science, economics, technology

Technologies as a natural background for additional changes in this so-called anonymous contribution, subsidies "10%" "from nowhere" in the infrastructural level of changes in initiatives that are also responsible for the background level for the neo DNA of a given field of application in/on the total or product level of influencing a given state status of a given process structure of our interest in transformation, environmental management of matter and life infrastructure.

This approach to the sources "from nowhere" means the sources of changes that interest us beyond the hitherto linear determination of causes.

Adding artificial/additional entrepreneurs to be reshuffled violates the current dependence and relationships system. Anonymous, i.e., from an uncountable/inconsistent, not yet created a source of supporting changes,

supporting the further shaping of the sources of directing neo/ evolution locally and totally – such neo-evolutionary initiatives, creations from nowhere, including a multi-matrix network of known and unknown/potential connections.

Pushing the initiative of changes to strengthen the security of one's own existential development trend in the always dynamic micro-macro cosmic trend lines

Throughout the project of dynamic applications in the vigorous competition of the environment, the so-called neutral and non-inert systems environments.

Just as caissons were used in the construction/foundation of the supernatural, then non-system Eiffel Tower...

Similarly, you can use supernatural initiatives to achieve seemingly unrealistic aspirations, projects, structures, and processes in attaining other goals than naturally calculated.

The greater the instrumentation, and tooling, i.e., precision, massiveness, and density, the greater the chances of interference with change and independence concerning the environmental background.

Therefore, stronger DNA, properties or efficiency, i.e., the new DNA or DAA, and then the new infrastructure of the material environment, and then the biological one.

These matrix applications in economics and politics will allow for direct and indirect changes in science and the economy.

The so-called independent creative decision-making platform from nowhere, from institutional propositions ...

To continue with the open strategy of neo evolution by economizing the process for breaking further barriers of life and matter...no more idle DNA but responsible DAA input.

CHAPTER EIGHTEEN

Infinite economics deepens a spectrum of DNA/DAA of physical processes, mutual technological transformations.

Infinite economics deepens a spectrum of DNA/DAA of physical processes, mutual technological transformations, the transformation of environmental infrastructure m, and further its factors characteristics Ev/Esk.

Without the economy that is establishing organization, goals, and means/tools of satisfying existential and developmental needs, technology would not have a basis for its development.

This specific network / multi-spectrum of dependence conditions of development is the basis of a particular network of neurons, experiences, ideas, and further transfer of competence in the direction, streamlining this process of hyper-spatial, hyper-temporal, hype-source transformation affecting the real, or rather more and more precise, determination of the time base and other factors deeper than the classical approach to time, matter, atom, DNA ... which could seem to us at a given level of consciousness, level, the potential of environmental infrastructure, its existence, as an untouched/unavailable level, point of development/ transformation/evolution/creation.

But this spectrum, this picture grid, of the environmental system is dynamic. It can increase or decrease our chances of development and existence. And it is addictive how we will actively perceive our participation in this process of life of matter. How are we going to live? Will we live? What will be our share, i.e., responsible construction of the foundations of this grid ... specific planetary engineering at a distance and depth of the structures of matter and life, where, here and now, in the organizational and economic field. We can remotely define the directions of the foundation, the building blocks of individual and collective participation in this process of exchange of matter and life of the micro and macro cosmos, which are still the suitable basis, a

guarantee of the stabilization of the ordinary vegetative / quasi-ecological economy that we deal with in most cases.

There is a need to exclude certain participation in thoughts and activities so that for the effectiveness of the development of infrastructure networks in the development, deepening, and consolidation of an increasingly better level of existential foundations.

That corresponds to the more individual, specific, anonymous, institutional, collective existential, and developmental needs on the increasingly conscious, effective tool – technical level of transmission, control, and knowledge.

The economy determines our and others' systems, decisions, tools, directions of development, attitudes to opportunities, potentials, and threats.

Correct relation of economic decision-making elements to physics, the biology of environmental infrastructure gives the correct dimension of the possibilities of economic, environmental, literally physical, and biological strategies.

Overlaying, and layering the network of initiatives/shares/ bricks is increasing opportunities, increasing infrastructure, creating your own background, bypassing, winding the system ... so that it runs fast but in the right infrastructural economic direction EskmDAA ... gives the suitable basis for a decent competition, overtaking other systems of life matter, based on their infrastructural environmental spectral meshes EvmDNA.

It could be used with decision-making elements of institutions, anonymous, autonomous, automatic, specific organizational and production tools, directly or indirectly, in voting, auctions, shares, bricks -DAA of this 10% share in the ordinary economy ...

The need for further specification, development of improvements, and consolidation of the infrastructure network to improve the controllability, properties, and efficiency of the environment of the matter of life, and this

the process in terms of economy, technology, and mentality.

Powerful support for High tech initiatives and hyperproduction, hyper infrastructure projects according to any tax reliefs for specific parameters of technological and production enterprises, public or private, according to the assumptions of the theory of absolute transformation of the environment, which is possible and applicable according to economic support for projects with appropriate production and technological parameters undertakings with hyperevolutionary features because only such will count on the evolutionary market of controlling infrastructure and, further, the properties of matter and life on the micro-macroeconomic and micro-macro-cosmic scale.

Projects and the system of separation of the so-called public and private, the references of which should be applied accordingly to the functioning of the mechanisms of the systems of the "natural" environment in us and around us ... in physics, biology ...

Such findings, working out a neo/evolution percentage of participation in every undertaking, process, and structure for neo-evolutionary goals ...

Every thought is an intention that will deeper infrastructurally support the direction of the depth of infrastructural intervention neo / environmental, neo/market, neo-economic, neo-productive, as a 10% share of hyper investment, core, a platform for further economic research, which will be the core of the following

projects, processes "m" infrastructure natural environment DNA and artificial DAA, their properties, efficiency Ev<Esk.

To be continued with the finding a hard and increasingly harder environmentally neo-evolutionary reference grid to the next scientific and economic endeavors.

CHAPTER NINETEEN

Neurons of the matter of life as the basis of a multiplanar approach to physical and biological phenomena in the EvmDNA

Neurons of the matter of life as the basis of a multiplanar approach to physical and biological phenomena in the EvmDNA <ESkmDAA process, where the factors influencing m infrastructure/ environment are the neural factors of the living matter.

Adding your own dimension, our own world of the infrastructural system, i.e., the new environment, taking part in the neo-evolution of systems, worlds where the economy and technology are to be the only boundary properties, the effectiveness of the micro-macro cosmos of the living matter. Artificial/natural neo-evolutionary pre / DAA neuron for new / environment, infrastructural life matter for ever newer EvmDNA

<EskmDAA.

Calmly, systematically but definitely, we are going to produce an infrastructural system, play dominoes, throw ideas, throw/propose, create new infrastructural systems, expressing our contribution to the structuring of the structure, and further to the effectiveness of the environment, and further efficiency in the implementation of our ideas for even more a better infrastructure, to break through the so-called natural physical and biological barriers, i.e., to own contribution and responsibility and rights in the share of equal rights and rights/laws analogous to the actual creative contribution to the creation of an environment of nature friendly to the matter of life, at a given existential and developmental level, and its further algorithmization that is the economic, technological instrumentalization of this process ...

Finding a hard and harder environmentally neo-evolutionary reference, foundation, and element of artificial infrastructure for the following scientific and economic endeavors.

Because a fatalistic science, a fatalistic economy based on the so-called fixed, specific foundations, the boundaries of nature with a separate, but together cooperation between science, economy through production, technology opens up new perspectives of the foundations of the development of existence, multidimensions the existing spaces of perception of nature and non-nature, which can also our and other world systems influence

the course of our cradle of life ...

hidden in the classic DNA atomistic, cellular ...

let's create such conditions for throwing seeds of ideas and technology

on each level of EskEvmDAADNA

You have to exit the play period for a given level

But although the potential and exercise can catapult our positions literally and figuratively in the "m" micro-macro neural infrastructure system ... this is how the virus catapults the behavior of neurons...

The need for powerful enterprises on the border of economic, technological / mega-production potentials because only such are responsible for working out their level position against the background of the infinite spectrum of evolutionary relations of the matter of life

... fun, but only in production

Fun gives the theme of the production DNASDAA driver, and it m dizzyingly accelerates, and it has the properties, the efficiency of EvEsk.

The Esk Ev is an economics combined fun of DAA and m

Let us give melodies, fluidity, and notes to this melody instrumentation process ...

It is juggling of the tossing of ideas, an ever-denser network of infrastructural undertakings of ideas,

Catching, working out, and directing one's own plan, a network of roads, which aims to implement ideas in order to produce as efficiently as possible, such as scattering tree seeds in the wind, and that the best infrastructure soil will fall ...

It is all about these economic technological breakers of biological-physical barriers and vice versa in the re-processed environment from a kind of different plane of this hyper-neural neo-environmental / pre-environmental decision-making system.

Economic and technological DAA controller as a neural grid of dependencies in a spectral tube of magnitude dependencies of matter and living systems.

The spectrum of triangles/funnels of the micro macrocosm, characteristic, wave, efficiency Esk ... about which the proposed models, charts I will draw later, and resembling classic images, cosmic, physical, biological models, defining the mechanism of life, the matter of evolution, the origin of terrestrial and extraterrestrial life, and further its development, controlling its development, its matter ... technology, economy, its multi-faceted nature

Pre-DNA mechanism of life of matter as a foundational basis

pre/neo evolution of the development of the fractions of the formation of life, matter as further levels of physics, biology, economics, the technology of materiality ... Is there existence of life beyond the so-called ours? – the above definition of matter and the life of its evolution indicates the neural nature of phenomena, and their dependencies – the question of who is stupid, i.e., more or less creative in this infrastructural structure of the basis for further control of the infrastructure of the natural environment ... its application, its rebirth, its transformation into a new multidimensional area nerve of economics, technology, fractures of connections and so on in the process of mutual building of foundations, the foundations of this process EvmDNA <Esk mDAA ...

A continuation in the compacting of this neural network of economic and technological dependencies and the potential for the existence and development of the matter of life.

CHAPTER TWENTY

EvmDNA as the primary environmental/infrastructural m1 degree is directly / indirectly affecting the further efficiency, organization/neutralism included in Esk and DAA.

And this is in the general support of hyper-natural transformation as a specific change of the environment for the achievement of new properties in further explorations, transformations, evolutions, and creations, but more according to our work, or rather the work of these systems, including ours, which by their commitment affect the course of the whole the process of transformation of the matter of life/life of matter, including our life of matter.

And this is the basis of true faith – creation/responsibility/ controllability at the so-called intelligent/neural levels – and not impotent dehumanization interpreter of the environment and infrastructure processes (physics/biology and economics – creative economy influences the interpretation, the course of all processes that are subject to creative, neo-evolutionary laws) – within the selective, neo-evolutionary, and hypernatural productive hyper-economy, which is the opposite of the sin of a fatalistic economy with a constant variable based on a single plane of transformational factors of matter and life.

Selective neo-evolutionary economics based on

neural DAA, which takes into account the baggage / next stage m and its neural DNA, is a strict production, ecological, economic policy, and – determining – a free market in physics, economy ...

Physical, biological theories ... including defining time as primitive relation to the infinite badness, the speed of the levels of environmental processing systems, in which we can infinitely increase our participation, appropriately redirecting the intricate structures of the structure, and further processes in the domino chain reaction processing. ... there is a fear of entering this sphere of action, which may end up looking – just like the impotent, insufficiently responsible, ambivalent ecological attitude of our civilization's economic and social ... processing, exploitation, production, science, and its application.

The belief in hyper economics and the sin of economics is the advantage of the selective neo-evolutionary economics over the evolutionarily dead economy of uncontrolled development ... cancer

DAA and DNA is a strict production policy, ecological, economic, and free-market of some physical and economic masses...

The energy of EskmDAA neo-evolution is the energy of creation infinitely more efficient than any blank mass of physical structures and processes because it builds up the further processing of any phenomena over the existing medium / infrastructural m. This mechanism of hyper neurological leverage should be activated, used, EskmDAAn = Esk (EskmDAA) > EskmDAA = Esk (EvmDNA) DAA> EvmDNA

It is hidden in anonymous, invisible initiatives in the classical sky of the matter and life spectrum but visible above the supernatural model ... always supernatural as ... the natural course of things

Micromacrocosmic physical-biological economic tool

It is about an increasingly productive and, at the same time, precise, massive initiative for digging out paths in this hyperevolutionary race to death and life.

It is a process, a domino phenomenon of the infinite energy of the environmental infrastructure, including our environment/ infrastructure.

We are its element, and we can continue, in cooperation with others, to develop our own trend of shaping infrastructure ...

properties.

It is about building locally, more precisely in a given direction of exploitation, intervention, transformation, infrastructural/ environmental and neo-infrastructure exploitation ..., infrastructural advantage.

So neo-environmental, neo-evolutionary

For more and more automation, i.e., the tool for unimaginable speed of the process of this EskmDAAn mechanism.

Controllability, setting controllability, and decision-making are the key to neo-evolutionary transformation and the general transformation processes in nature and beyond nature.

In the mass processes of the mechanism – the mass nature of changes

The lack of this element does not justify the effects; conversely, its presence determines the speed of this process ... Massiveness alone is not enough. Still, its absence is also insufficient in determining the directions of the rates of physical and biological transformations, exceeding the scale of the existing model measures ...

Everything is possible – it is enough to

EskmDAA a balance of initiatives always of a mass character.

The environment and its change, change its properties and potential; we can be active or passive shareholders, but only on a large scale of decisions, actions, processing, neostructuring ...

It is about anticipating the broadly understood evolution of the processes of matter and life in occupying the infrastructural, environmental space ... defining being, or rather its chances by shaping its properties in this way ...

We, our science, our economy are to take this into account in their models, as models more active in ... fighting the resistance of nature, fighting the cosmos, or rather the resistance against the so-called mentally imposed and "actually" laws of nature at given levels of exploitation ... breathing, life, civilization, but also in combating the primitive economy, in favor of more developmental, responsible concepts in the dynamic cosmos of matter and life systems, at the hyper macroeconomic level, hyper scientific, hyper cosmic, from and for hyper matter and life ... beyond that already known natural, neonatal, own natural, able to cooperate, compete with these natural competitors to an increasingly efficient, organized ... algorithmic degree.

To be continued with the Lego blocks without horizon of the universe of life and matter...

CHAPTER TWENTY-ONE

Charts of the breakthrough of the spectral surface, interpretation, and application of the mechanisms of changes to the current trends.

Charts of the breakthrough of the spectral surface, interpretation, and application of the mechanisms of changes to the current trends in the processes of matter and life structures.

Such propositions, divine inspirations, and miracles from nowhere, i.e., other planes of applying the adjustment to the same pattern, were finally approved in point-linear terms.

It is about the mechanical, technological, economic, and organizational influence on increasing the probability potential in participation in neo/evolution, i.e., wonders of nature – over a plane, over three dimensional.

Simply everything, with the help of a network of feedback phenomena, processes, and structures, we can influence the change of their properties, their further existence, and the evolution in the macro-micro cosmic sense, matter, and life ...

During the holiday period, in the so-called free time, I will suggest loose graphs, models that may turn out to be important in a more detailed description of the non-linear spectrometer of phenomena, processes that can be used for a multi-directional, multi-source process, the so-called natural effect – described in some 2/3 dimension as well as supernatural, going beyond the hitherto areas of natural, artificial activity, in evolution, the creation of all processes, structures that have a more or less direct impact on our environment of life of matter.

It is nothing less than a project, a model of a new methodology, new multi-equilibrium imaging in determining and influencing physical and biological phenomena of the micro-macro cosmos, where the economic and technological factors will be the limit, and at the same time, the infinite potential of further shaping, changing, inverting, rearranging, proceeding. of existential and developmental relations in the field, atom, DNA, technology, production, ecology ... Such a hyper model of DNA and atom, technology, and economy in one

It is a mega-open project intended to be an additional impetus to the city

in a more effective, more multifaceted mechanism, an algorithmic description of the terms, phenomena, and processes hitherto, in every area of life, human work, and his nature in him and beyond.

Over natural, over temporal (another definition – more playful illustration of time/processes) over spatially, for a better understanding and laws of phenomena. Spectral multidimensional description, and at the same time, after its training, influencing the re-directing of phenomena, creating your own phenomena, which with this multi-spectral algorithm are to be furthermore efficiently, autonomously / = tool-related and applied.

In the application of this multi-living model, the tools of change need cooperation, your further ideas to implement your own and other plans, including a neo-algorithmic non-linear description of phenomena, shaping these phenomena ... because we are part of them, the question of how efficient, responsible, worthy of nature, god.

To be continued with the charts of neoevolution, a miracle of neo/creation.

CHAPTER TWENTY-TWO

Nomatter(=processing) makes the (new)matter as the creation of its own nature/infrastructure for creating new processes of matter/ life structures.

Nomatter(=processing) makes the (new)matter as the creation of its own nature/infrastructure for creating new processes of matter/ life structures – the essence of neo-creation and evolution- of going out from nature...

Brainstorm or eureka about the no/matter of/and life and so its neo/properties...

Always ever newer infrastructure m – as new nucleus as newer property of/by new particle for new directions/dimension of physics and biology...

Increasing the potential indefinitely, infinitely quickly thanks to a new interpretation, a new model of matter, mass, ... according to this universal nucleus-neurons model of the EskmDAA graph ... indefinitely strengthening the DNA structures of... telomeres..., other constructive atoms, cells ... oncology (the cell does not deteriorate aging), atomistic (infinite process, structural decks in EvmDNA/Emc2 ... according to <EskmDAAn model.

This algorithmic graph of modeling, materializing, and neo-materialization of phenomena, processes, where this multi-plane model of matter indicates the so-called immaterial factors of further processing of matter, whereas if no longer material resources, processes determine and influence the shape of this visible matter, the matter of life.

The following processes and modeling illustrate differently and define matter as a means/environment/infrastructure for further shaping the plasticity of natural and unnatural processes. Unnatural, that is, away from the current image and perception of the environment. Derive formulas, models,

and charts more and more dynamically, practically a picture of infrastructure, neo infrastructure of the environment, its material foundations, neo infrastructure, neo-materialization of the foundations of environments, neoinfrastructure of the environment – more practical computational possibilities, program predictions, influencing processes ... just like playing a game bowling, where more and more precisely, not only on the level of the Mendeleev table, defining the bowling game, algorithmization, structuring, and further steering, evaluation, drawing, steering, infrastructure (on our part) of the environment ... matter of time, a matter of energy, matter of space, where the existing models will be only a small component for further transformations, evaluation of scientific, economic, political, civilization, ecological, religious, matter and / life models of the micro, macro cosmos.

Let's try to get closer from every possible point of view to this neomaterial model of the spectrum of matter and the life of the micro-cosmos with the help of this not one-zero but EskmDAA> EvmDNA ... where the m in EskmDAA covers the entire formula EvmDNA ... the same applies to EskmDAA1> EskmDAA, EskmDAA2> EskmDAA1, etc...

It is like a further continuation of building the development of DAA awareness. Then the construction of m ... – economic, technological, futuristic, civilization, scientific intelligence – their combination or lack multiplies the effects, models, and strategies...

If there are no scientific arguments, a breakthrough requires economic and organizational political arguments and vice versa.

Politics with absolute nature, voting, trending, "buying" votes ...

It changes, adds dimensionality to infinity ... EskmDAAn> is this model of a breakthrough; it is this addition, a constitutional amendment to the matter and / of life.

There is no justification for any status quo of matter's life. Any surrender is a sin of omission, a scientific, economic, social, political sin, a sin of complicity and suicide, murder ...

Here, this activity should be appropriately implemented in this process of control, influencing the processes of the structure of matter and life – but this work is available according to the needs of ordering and directing.

This direction is the basis of the EskmDAA formula as the basis for the construction of the next generation materialization according to this work by EskmDAA...

Projection of the building material's own trend

3d, Mendeleev's own artificial table of materials to create a separate infrastructure of processes and structures in the management, intervention, and creation of own micro and macro transformations of the cosmos, own micro-macro of the cosmos. A model, a chart covering entire structures, processes as only a material/point to be processed, as well as its ... omitting, going your own way, creating, learning your own creation, revolution.

Above natural, unnatural, i.e., not the current one – a fluid dividing line – it is about moving away from the existing models and processes in order to approach them, or take over, omit, understand ... the nature of creation, ...

Create your own nature/infrastructure for creating processes of matter/life structures – the essence of neo-creation and evolution-by gradually ... moving away according to EskmDAAn> EskmDAA2> EskmDAA1> EvmDNA, where m from the more recent formula includes everything in the previous formula/procedure. It is a pattern for reactions,

explosions, creations, creating evolution in the so-called process, natural and unnatural, i.e., different from the natural, more and more divergent ... treating us as m, and we are not natural to treat this nature as progressing in our own drive to life, survival, and development.

This is the sense of seeing the world, one's laws, the place of nature, "nature" of life and matter in every private, public, scientific, economic, religious activity ...

... that's how you can predict the future

To shape, explain the future, ... the past with more and more precision, just like with the ever-increasing number of bowling pins and machines targeting them with a ball ... better and better Esk... than the previous Ev...

So nomatter makes matter.

Making situation make new matter EskmDAA from m by m = EvmDNA

Whatever situation creates new matter, the basis for subsequent situations, and so on ...

The new matter of new/artificial Mendeleev's infinite array by EskmDAAn> EvmDNA ...

The non-material process model of neo/matter, its neo/evolution (not revolution — neoevolution is a more potent aspect of changes).

Moving the boundary (of potential), the plane of illusion and reality in the perception and control of the environment.

Moving the boundary (of potential), the plane of illusion and reality in the perception and control of the environment, and simultaneously the boundary of the possibility of influence, in applying the structure model only in the process approach.

Illusory and real actions at a given level, a given plane/area, a space of technological and evolutionary niche, which determines the continuity of the process and not the structural (temporary) statics as the key to understanding evolution, revolution, neo-evolution, devolution in the creation of the phenomena of matter and the life of the micro-macro cosmos on at any level of human and other system activity.

Before Copernican, Copernican, after Copernican, before Einstein, Einstein, after Einstein, and other discoverers, descriptions, and models have always, has, and will be applicable at a given technological and application level.

The more thought-practical methods, proposals, and attempts to act, possibly directional, the greater the base, then the thoughts of action, purposeful actions, effective actions, but it is to be of a maximum character to the resistance to feeling the directions of influences, transformational / process-neo/material, neo-infrastructure, even such for no reason, but just catching up with growth and risk, not with the indolence/laziness of natural subordination

that is, it is about an ever-increasing investment, with an ever-increasing system of backing up this initiative ... business, growth, because that's how other systems are developing in the evolution of the environment inside and outside of us.

Application in business, politics, economy, science, economics, ecology is a mechanism ... with more and more

powerful acceleration, intervention ... but with an increasingly larger security system – and at the same time, and therefore the expansion of the neo-infrastructure, neo-environmental – neo-evolutionary base, if it does not want to be eaten by other systems, which are involuntarily subject to actions, influences only / already at the expense of our living space, matter, the potential of used matter, influencing, developed in the process on each other, i.e., the expanded base of growth and security.

As with Pharaoh – not democratically – so currently undemocratic scientifically dogmatically – and it is about a bottomless, planeless, boundless change of material foundations, or rather process-based "structures = models" of the environment ... in direct / indirect participation in creation, and not or taking responsibility for the totality of enterprises (ours and ... other systems) on the border of security, comofort border, security buffer, as well as a buffer, infrastructure facilities to be a soft / flexible platform for enterprises according to the choice of tools from this platform .. (no blaming the blame, responsibility on the pharaoh / nature / god ..., because in politicians, scientists, entrepreneurs there is this joystick, a tool for securing theoretical, practical decisions ... organizing infrastructure on this possibly deep processes of "matter =" model ", to the implementation of our choices, dreams – the more extensive / bypassed system m the better ... flexible, responsible, we can effectively develop what is the basis of processization, existence, life on the physical, economic, ecological, boiological, medical, civil, political, religious level.

Nonmatter processing, creating evoluting new matter that makes new (non) "matter" processing that makes new "matter" .... for the "matter" properties changes

that is

it is no longer a material treatment of the environment, but a process-like treatment of them, if we want to go beyond the status quo of the life of matter ...

Processing instead of structuring defines the properties that control the "material" (no longer a classic approach to the material ... which can further control important decisions, processes, "material" .. processes of forces of the so-called matter, matter or life ... the dynamics of materialism ... over tables Mendeleev's structure of matter, above the broadly understood level of pre-Copernican thinking about matter and evolution in any natural and unnatural activity.

Such a steady trade, i.e., processing as an actual materialization of decisions and ... evolution, "active commercial cash" as the basis of existence, development, strengthening, improvement, makes the effectiveness of a given process of "matter" more flexible. Farming, improving the process of economization of matter turnover, improving the matter, increasing plasticity, further flexibility, higher infrastructures, higher serviceability, better potential properties of this matter, or rather the processing of structures .... economic improvement but ... such a physical plasma .... creatively competing with the neo-infrastructure constants of the environment – no longer a permanent environment, a permanent status quo, but infrastructure in active participation.

It is about treating structures more deeply because if we do not, other dynamic systems have changed. It will change these so-called constantly (in fact, it is the surface coating/outline/tool/

processing resistance limit) of the structures and their properties ... a fuller open image before procedural nuances, contributing to all structures/systems and their properties.

Obviously, this dynamic alternative approach to the state of affairs changes the system of forces and assumptions in terms of science, economy, politics, civilization, society, religion, and their dynamic response provides the basis for the changes ... the question of their application.

The question of applying the right type of algorithm, the tools of change we and other systems have always used,

will apply.

We can and are fully responsible for changes,

for changes to an increasing extent. The issue of developing these foundations for processing the basis of structural materiality, infrastructural dynamics of approach to the environment in us and beyond ... and taking the position of a passive consumer, more or less, but a co-producer on the preliminary scale of ... rights and responsibility ...

Let us treat the environment as our own body. Environment-process in which we participate passively or actively, as actively, responsibly, and legally- literally establishing the properties of matter and / of life.

To be continued with the neo/evolution against d/evolution.

CHAPTER TWENTY-FIVE

New language, new grammar, where active process relations will be used, which are to be the proper factor(DNA) of matter/life.

New language, new grammar, where active process relations will be used, which are to be the proper factor, property, properties, plasticity, the structure of matter, mass, energy, life ...

Yes, a new grammar, a new language, literally. Not only a new language of physics, biology, medicine, cosmonautics, and economy but a literally new language, a new mentality of thinking, acting, using the right words and phrases to act, to overcome the barriers of seemingly supercosmic impossibilities ...

We will penetrate the structures more and more actively, their fundamental, process-like nature, not so much ... just mass, energy, gravitational, light, static ...

A not-so-devolutionary look at the so-called structural inevitability of processes, in which, however, this processability may reverse this structure – processivity with an infinite micro, macrocosmic potential according to Eskn (Esk1 (EvmDNA = m1) DAA1) DAAn.

This is my (one of many ways, planes, spaces of the net of life and matter of... nature) view of our world. My proposal of a tool, a method of entering the mechanism of the environment, environmental engineering is not so classically ecological/devolutionary attitude, but hyper ecological – neo-evolutionary – that is, the dynamic economics of the co-evolution of the environment.

The question of how many conclusions and suggestions will be made.

Phenomena of action, not just matter ...

We do not need to prove our position or history, but actual current actions define and determine situations and

phenomena that would come – not to have, but to be, it reevaluates the grammar of language between us, any activity and phenomenon .. . Creation/creativity releases the dark matter, .. that the earth is no longer flat, and other dogmas are also not so "square".

Matter means where, how is

action, processing determines any phenomena, show the over the flat picture of matter, masse, energy ...

It is the addition of further technological and evolutionary niches to emerge from the devolutionary imaging of phenomena, processes, and structures.

Exiting technological inertia for ... Emc2

Conviction and truth at a given technological level always create illusions of the limits of possibilities ...

But everything can be implemented and changed.

The issue of constant search for niches to achieve directions of change, transformations ... Light, time, etc.

So not evolutionary inertia, not maintaining the evolutionary status quo, light ... technological status quo, because it is, in fact, devolution, depriving oneself and others of rights and laws ... to further development in infinite potential, often invisible beyond our technological, preceptive the sky of the world of micro-macro cosmos of matter and life, a matter of life, a life of matter...

Aevolutionary or devolution and evolution and neo-evolution and revolution

This is best described from a niche perspective.

We are talking about a neo-evolutionary niche, i.e., the possibility of redirecting changes in nature inside and outside of us to a given plane.

Devolution niche, i.e., niche deficiencies, lack of initiatives, exploratory, exploitative, economic, and mental deficiencies in the possibility of redirection

Scientific, economic ...

Devolution is also inertia, evolutionary laziness, the antithesis of neoevolution, which is against the potentials, threats to the evolution of matter, and the life of the micro-macro cosmos.

EvmDNA evolutionary inertia, but the next one will always be inertia; you must constantly maneuver, search, and expand the horizons / sky-high world of events to come out of a constant collapse – EskmDAAn neoevolution is the obvious continuator of life, i.e., anti-inertia ...

This inertia, this d/evolution, is an attempt to materialize, limit, and freeze.

An attempt to freeze ... knowledge, phenomena, processes, their mechanism, freezing cognition or learning merely a frozen static image of a model, it is two-dimensional, not dynamic 3/4 dimensionality .. errors, illusions of science and economics ...

Processing spectrum with upward effect

mass spectrum of energy and other structures for creating objects, their processing as a tangible border image, a measure of reception, but also feedback control, influence ... like the light spectrum of emerging and dying stars ... their infinite control potential, the feasibility of controlling these and other processes on a scale unimaginable by us today, as 1000 years ago it was unbelievable to send space ships, production in millions of cars …

So known spectral models, patterns, and new "non-natural" artificially produced, produced … in

a multifaceted (infinite size of the possibility of process matter) process network as a spectrum of the area of a given phenomenon of an object, mass material in evolution, implementation of any trends, goals regardless of the infinite material core, based chiefly on process dependencies, tendencies, initiatives … neo-evolutionary, neo-evolutionary / natural …

Here we are talking about the new grammar language of the words physics, religion, politics, economy, … literature, from the position of term, expression, word to have to process in a constant trend of description and change of phenomena, the phenomenal character, the foundation of matter and life.

This is the description of the world turned upside down …

Such a new path to reaching your goals, as long as they are not fatalistic. Each consistently developed path, course, and approach of development in economics and science has given and will continue to provide relevant results. Here, in this case, we are talking about the effects of multi-global, multi-disciplinary, total, and … in every smaller civilization enterprise …

CHAPTER TWENTY-SIX

Disenchantment/ neocreation of structures(processes) of the micro-macrocosm, i.e., the awareness of the economics of neo technology/matter production.

Disenchantment/neocreation of structures (processes) of the micro-macrocosm, i.e., the awareness of the economics of neo technology/matter production in the infinite configuration of materialization restructuring.

Yes. Awareness of processing as a guiding factor of the growth of control, reproduction, neoproduction, technological neo-constructions pyramid/evolution (abnegation of the process is devolution) of the existing structures – procuring/influencing the flow of changes, the tendencies of matter and life to an indefinite degree – their structural, material, as if their plasticity and their procedural basis.

There is always a further hyper-technological response to any pattern of a physical, biological model ... it is about conscious, more effective, productive redirection, breaking ... the laws of gravity – their deeper structural / process factors, or rather technological ones. It is about productive, economical (as if a specific, evading another path, or maybe one shortcut) in competition with visible and invisible systems, factors by achieving an increasingly higher self-construction system of solutions, creating your own DNA (DAA) of further neo-evolutionary material, or rather a process which is always infinitely flexible in nature. The question of the organization of the pyramid of events – structural processes/ function in the atomic and other matters ... the chances of success lie in the consistent enforcement of this process at every possible level as the final source and effects in this constant, infinite flow of neo/materialization... of matter.

This peculiar accelerator of money calving of infinite potential and options of results at every stage, level, and place, and subject, arrangements, models, sizes, and levels of processing as the primary and infinite source resembling its mechanism of operation in the image of the model of the process of a magnet, but also its the

endless possibility of its evolution as well as the evolution of each element of matter and its properties.

Yes. Shaping the evolution of processes, their plasticity, palpability, limiting, shifting the boundaries of a given level

it is a scientific, economic, technological ... religious, and moral stimulus for expansion –

otherwise collapsing – its sustain and development in any direction ...

more or less specific path of matter technology/matter processes

opening the sky of events ... is the reverse of the singularity of creation

So here we are talking about a new grammar language of physics, religion, politics, economy, ... literature, from the position of word to have to process in a constant trend of description and change of phenomena, the phenomenal character, the foundation of matter and life.

This is the description of the world turned upside down ...

Such a new path to reach your goals, whatever they are, as long as they are not fatalistic. Each consistently developed way of development in economics and science has given and will continue to provide appropriate results. Here, in this case, we are talking about the effects of multi-global, multi-disciplinary, total, and ... in every smaller civilization enterprise ...

The effectiveness of the process of processing, i.e., material and qualitative manipulation of qualitative manipulation in a continuous process, image, spectrum, depends on taking into account one's own and other process development factors, which always have infinite direct and indirect repercussions – without stopping the development, on maintaining development ... in the pulps of other systems .. our internal systems ... parasites ...

There is no materiality in the masses...

simultaneous indirect processing in past time and constitutional gravitational guarantees is the title of the existing processing – not materiality to only derivative procedures

this approach to the materiality of the mass of properties, the quality of possibilities, has an overwhelming impact on further scientific, economic, civilization, social, and religious processes ... of literature

formation and effectiveness of the nature of participation in this process

Multidisciplinary influence on the work foundations of matter and efficiency. Phenomena with these related processes as the basis for changing the phenomena of matter in physics and biology of neo-evolution of shaping matter employing processing in a chain reaction in every field.

Processing, continuation as the basis for materializing its properties, influencing its further neo / evolutionary contributions and effects of processing neo-evolutionary materialization, for more effective participation of variable matter in this process.

Economics, social and family goals ... but not only between human purposes but also above all with the systems of nature, which give tangibility to the boundary of the material of processing – Processing = economy – never rigid, neo system, creating material neoinvestments according to the programs of economic entrepreneurs / = ecological is the mathematical result of relations with nature in a direct collision and coexistence.

The level of the illusion of reality based on knowledge of technology ... sufficient, but the technological approach

raises the level of ... science, a religion of science, the science of religion too ...

technology etc., according to EskmDAAn, i.e., Emc2, Copernicus, Bohr is at some level but never sufficient for further development ... it is enough but not final in further existence, expansion of the matter of life, a life of matter(=process)!

There is no evolution if there are no multi-faceted, infinite size and speed systems of systematizing systems and neo-evolutionary speed – the revolution works on the same level, and devolution is ignorance, the so-called politics, religion, science of the approach of helplessness. Some kind of permanent barrier in the relations of evolutionary processes – not going beyond the "Emc2" of evolution, but rather stagnation and even collapse.

And it is about literally and figuratively breaking the sky not of matter but of events, which are the foundation for the materialization of its evolution

... everyone is rich – let's create, create the foundations, mental, psychological, technological, economic, scientific foundations for their expansion of potential, with infinite potential.

This is divinity, the creativity of this process, which we are able to change with our hands, with tools, endlessly, with ever greater speed to infinity, and not praying passively to the course – praying as inference in an ongoing process of events ...

The church can support this in realizing the divine creative image of the development of the world and not a collapse or devolution.

It is quite okay for contemporary models and patterns, but let's try to open them and neo-materialize, new planes of the basis of the pattern potential, even in opposition to apparently closed facts. Let's rehearse; let's practice it.

Processing spectrum with upward effect

the spectrum of "a/syenergy of other so-called permanent phenomena, objects, and yet the spectrum of processing is the spectrum of the birth of new niches ...

The need to change, genesis, decipher the words of the grammar of meaningfulness of the object of a thing, the boundary of the process barrier, the infinite potential of participation in it ... the specific meaning of life, the mechanism (materialization = processing) of its existence, maintenance, development.

Processing – means the infinite "material" potential in every case. Its infinite plasticity in attracting technological, operational, and engineering sources, a kind of increasing or decreasing the potential/possibility of data, but of endless size and material development, in a multi-plane, multi-spatial approach.

To be continued with the grammar awareness, the (artificial) intelligence of processing, DNA / DAA perception, its technology, and the economics of this technology.

CHAPTER TWENTY-SEVEN

Going towards a new model, the formula of matter – neoevolution/ creation/action of matter/=value.

We may be like lawn-eating cows or EvmDNA, but let's put it further in not d / evolution but neo / evolutionary Esk (EvmDNA) DAA brackets.

Without looking down on nature ... matter, evolution becomes devolution. It is the lack of supra-gravitational action or negative dematerialization... a/action, positive or negative amatter ... that is, responsibility and not evolutionary nihilism – scientifically and economically ...

A given intelligence / artificial is always based on the competence awareness of a given link, expansion of this link, improving the processing of structures ...

Involuntary awareness of intelligence as automatic and maybe also a worked out contribution, de/neo/materialization effect in a continuous, infinite processing cycle, which is also a derivative of this intelligence.

– effectiveness of the reaction, experienced, developed response with the help of the developed/experienced environment

These are our additional process, environmental hyper artificial or "natural" synapses as well as adding more and more robust structures – controllable efficiency processes ... man and stone, hammer, and in the further evolution, human neoevolution and giant control production modules becoming more and more powerful, further, deeper environment / micro-macro systems of the cosmos, which are programmed, further processed, built up from it, creates further control / "hammer" facilities further than the hammer range of older production procedures in infinite areas of the environment, but with our unlimited potential to plasticize the phenomena of this environment, increasingly more expansive areas of the environment , including the area of planets, cells, atoms, quarks, phons, stars, galaxies, entire cosmic assemblies, macrocosmic, microcosmic matter and life, where their underlying factor of existence and development, the direction of existence and the direction of growth/action in a given area is always these and other processing.

Gradually smuggled in every take, the word procession as the fundamental changes and shapes the environment, shaping the cosmos – the level of involvement affects the chances and sizes of transforming the structures of matter and life, materializing life, or rather the processing of matter and life.

Processing, joining in the processing of any structures, phenomena is taking over, or rather deserving it, developing – creating a background, investing in the background/=envirovment, maturing, consciously taking over competences/control / DAA, ... not throwing the blame, the burden on fate changes, nature (absolute models) and god.

Develop competence/awareness of competencies, not

shifting to God nature ... these competencies, i.e., a materialization of own property of the matter of processes ... from the development of these competencies...

In understanding one's competence and responsibility position, one can take a closer look at the two-track definition of evolution.

Devolution and neo-evolution as the primary exponent of taking responsibility ... that is, the birth of divinity from the cradle in an effective, profound, economic, scientific .. political, social, religious, moral outlook on constant, dynamic evolutionary changes, where passivity is synonymous with devolution in existence, development, on the active impact on the sources and resources of growth/action, as a neo-evolutionary step in emerging from evolutionary/=economic/=   hyperecological   environments, gravitational phenomena – literally and figuratively.

Awareness, intelligence, perception of DAA

technology production m + efficiency Esk = infinity

shaping evolution production is controlling programming with the structures of matter

production control, control of processes, structures – processes, structures of matter, which is to further affect further efficiency – the lack of this process is devolution in relation to the surrounding processes and phenomena of matter and life ... economics, science ... politics ...

Control and acceleration of processes, and thus structures that are always derived from processes, whether natural or unnatural – is possible to an infinite degree. It is a matter of economization, where in ... in the growth process, i.e., change this unprocessed invisible into a processed visible effect/matter, which, after reaching the visible range, can be further exploited, ... mastered ... to an unfinished degree

all scientific and economic projects have so far been devastating mentally and technologically

by pending each achievement of fatalistic limits of possibility and impossibility trending/evolution lacking this hyperevolutionary addition

... the economy also needs this endless catapult in development for each area of life/matter, mutually supporting each other structurally, hyper / productively.

To be continued in this puzzle, the intersection of unknown factors, words that are not always related to each other at first glance but that inspire widespread initiatives.

CHAPTER TWENTY-EIGHT

Action => process = < matter > (mass, phenomenon, ... energy)

Action = matter, the size of which, the advancement of the action/process appropriately outlines the factual background and rigid material foundations .. which fundamentally, totally changes the economic, civilization, and scientific background of humanity, any civilization, and life. ..any matter and phenomena.

Each action/process/phenomenon +/x matter (any issue/any phenomena) = creation of ... new matter/value, as always the next degree of its change, d/neo/evolution, in potential and actual degree of acceleration, an infinite, dependent on these layers of these interventions interested in the directions and intensity of the changes.

Creating, showing, participating, working, engaging in the model of the mechanism of changes, the speed of changes, the depth of changes, including the material of the environment, developing the material of the environment, and at the same time, tools, infrastructure, on which you can further implement, produce further paths, tools, platforms for this evolution, creation, i.e., participation, already inscribed involvement in the direct, more and more direct in the structures, structuring of existence, in the breathing of the environment in terms of direct economic, physical, biological ...

Let us take a look at the attached sketch of the diagram illustrating the specific spatial jump of one matter to another depending on the intervening side (also the so-called natural side), procession, initiating action, process, phenomena in a given scope or over a given matter, situation, occurrences ...

It is an open project for further interpretation, expansion, and implementation in each chosen activity direction, human matter structure, and the environment.

I am throwing this model, idea, striving not to impose the course of thought, albeit each initiative has its own individual vision, point of view, the question of the effectiveness of this idea in application ...

Let us give ourselves time to reflect. To apply my waiting sketches and the ones that follow.

Let's be calm. Take it easy. We are still in the summer season ...but climatically sweltering season...

To be continued with/for the taking right place in the further expansion, economization of the sketch of the production model of materialization <the processing of each structure, and further its properties in the graph of the infinite acceleration of creating changes in a negative or positive direction (d/neo/evolution).

CHAPTER TWENTY-NINE

The economics of infinity, i.e., instrumentation, industrialization of a mass of the matter of life. A new science of economy, a new science.

The economics of infinity, i.e., instrumentation, industrialization of a mass of the matter of life. A new science of economy, a new science. Economy, science, ecology, biology, physics of infinity(of any process/action/= of matter/energy).

Dredging, deprecating a solid by action/decision means there is no more of this geometric solid (there is also no solid in strategic thought), i.e., neo-evolution, neo/material, as an active basis for controlling all phenomena, processes, and structures in each case in degrees and scale to infinity, accelerating this process to the infinite, ahead of all known micro-cosmic phenomena in scientific and economic terms, to a practically hyper absolute degree.

Just as Einstein made his inference from previous perceptions, patterns, and models, so and now, continuing, we can see that the processing of phenomena (matter) is the foundation of "structures" or rather an infinite set of processes, a network of processes/=operations that addictive, burdensome, determine the tendencies of the environment in us and outside. We can also add our added value, as it used to be in natural and civilization processes.

It is the resource management of this infinite because it is based on the action of the potential of matter.

Production, instrumentation, instrumentation, automation, ever deeper, further manipulations, instrumentation = automation as increasing and directing processes, phenomena that are to be, are always the new, worse or better (from the point of view of interests) basis for the materialization of matter.

We have no way out in this "gravitationally" very intense, non-static environment of matter. If we want to develop or at least secure a niche for our existence without looking at the "roller/fate" of permanent change.

Without a responsible, aggressive environmental and ecological superstructure, we have no chance of any grace.

We must become billionaires – each individually – in the economic power, in the productive power/ability to influence these processes.

How to make ourselves billionaires of matter and life in terms of economy, civilization, evolution, physical, biological, astronomical, and cosmonautical in achieving in these topics cost-effective production scale, really mass in this topic, goals?

... Self-mutating incubators (on the ground, ... on the moon, on Mars), production, tools generating new mechanical autonomous economic tools are more responsible for participation in the processes we influence, but nowadays not so responsibly, just more so with a slip by the margin of error ... caused by strategic, life, material nihilism – in advance, encompassing some invisible dome, the limits of impossibility -stability, or rather the collapsing of material, life – such a kind of subjection, pushing on a god, pharaoh – that is, from the assumption of one's own irresponsibility ...

Contempt for ourselves is contempt for nature as for some permanent monster, in which we nevertheless have all the rights to participate as a part. – Yes, it's just ritual / (un) scientific / (un) economic suicide with no participation or omission.

Some plans are basically suicidal plans from the assumption of religion ... science (modern religion) of some solid model, that is religion of the spherical dome of the world and life, which somehow does not go beyond certain "gravitational" boundaries. It is about building a new system of values, and properties based on such a peculiar creation of an evolutionary soup of matter through actions, actions, possibly directed, sometimes erroneous, but developed in relation to the absolute negative phenomena, negative ones that double the foundations of further existence.

_ Powerful forces are directed; they are rolled by the structural results of the process, so that only a change of their basis, for the mismanagement in their processing, not directly related at first glance to the existing scientific and economic, logical, but only here and now activities

Each such action contributes as each drop from a comet, a meteor contributes to the fundamental metabolism and life literally and figuratively at every level. The greater the contribution, the greater the chances (input = action> / = matter and life = life of matter / matter of life

the more ... errors, exercises, training, the greater the chances of discovering, directing, finding, producing, creating, neo / evolving foundations other than competing living systems of matter and life in us and beyond – including genetics / deeper basis of gravitational loads on the matter and life in the areas of our classic scientific, economic ... social, civilization activity

society ... its active activities

Action group action expresses our identity, ... materiality as a further basis EvmDNA <Eskm1DAA (m1 = EvmDNA)

A neo/new materialism, neo ecology, neo biology ... neo medicine..neo cosmonautics/astronautics .... by action always > matter.

solid < tool – economics

The fact that matter is not obliterated but produced, automated is the basis of the economics of the infinity of evolution, matter into a tool that becomes matter, a solid not to be missed, but to a flint

......

_ Powerful external / internal forces – action> matter –

they influence, direct, and determine the process, and structural results, so that only the change of their basis, for modification in their processing, not directly related at first glance with the existing scientific and economic, logical, but only here and now activities

Each such action contributes as each drop from a comet, a meteor contributes to a fundamental transformation of matter and life literally and figuratively on every level. The greater the contribution, the greater the chances (input = action> / = matter and life = life of matter/matter of life

the more ... errors, exercises, and training the greater the chances of discovering, directing, finding, producing, creating, neo / evolving foundations other than competing living systems of matter and life in us and beyond – including genetics / deeper basis of gravitational loads of matter and life in the scope of our classic scientific, economic ... social, civilization activity

society ... its activities

Action group action expresses our identity ... materiality as a further basis of the matter of life, i.e., activity and materialism, as a material attitude, the foundation of material properties ... of matter/values, ... life, but without the passive, fatalistic, classically scientific ... physical, medical, biological, chemical, but economic look at nature and god! – EvmDNA <Eskm1DAA (m1 = EvmDNA)

Yes, it is already neo materialism, neo ecology, neo biology ... neo medicine..neo cosmonautics/neo-astronautics .... by always action> matter affects the properties of matter, i.e., the dynamics of the infinite layers of the foundation, the potential of matter(mass), any matter/energy/=/<life.

Any matter as a solid, as a future tool, a material for a new material basis, a new solid for the further continuation of the transformation -> a block, a tool, a better block, a better tool, an even better block, an even better tool, for further processing and more and more productive, i.e., economical on an infinite scale, i.e., neo / evolutionary, more efficiently with an increasing share, responsibility on our part if we sincerely want to count on anything.

This is the economics of infinity. Yes, a whole new section of eco-science and science, science and economics. New economics, new science.

The non-blocking of matter, its instrumentation, production, and automation is the basis of the economy of the infinity of evolution, matter into a tool that becomes matter, an impenetrable lump, speaking from flint ... to a rocket, to the moon, to stars ...

CHAPTER THIRTY

From scientific, economic, and religious nihilism to the strategy of neo-evolution, neo-creation of the life matter environment.

From scientific, economic, and religious nihilism to the strategy of neo-evolution, neo-creation of the life matter environment, as a new civilizational, existential basis for the entire micro and macro environment of the cosmos.

Continuation in the relentless struggle with scientific ..., economic, astronomical, medical evolutionary, and creative backwardness. Provocative. Leaky? Yes, for those peculiar doctors of civilization death who look at the so-called broadly understood collapse of the processes of the evolution of matter and the life of the micro and macro cosmos.

Of course, it is not about rebellion itself, but simply constructing a strategy for reviving one's position in directing, creating the environment in us and outside, not looking at fatalistic looks, reflexes, a system of thinking and acting based on a block – a rigid/ structural view of the world, its process, infinite potential basis.

New science, new economics, because new nomenclature, new grammar, new thinking, a new term, active determination of non-static phenomena, structures, processes, a new grammar of science and economics, .. economics of science, the economics of matter and life. A science worthy of absorbing space on a mass scale. Creation of matter and life. Creation ... leaving the wall of civilization and mental collapse and not burying ourselves alive, pseudo-ecological and pseudo-civilization.

Active, process-based description of the world. Not static,

dynamic, changing the grammar of the activation of matter and life – from the past to the future now.

Totally changing, affecting economic, scientific, mental, social, political, and religious consequences.

The foundations, places in the environment, which were, are to be and are our actions, initiatives, and initiatives along with other actions/activities of other sources, dynamically to a potentially infinite degree, as a result of a domino chain reaction.

Paradise ... it is building further models, structures, expansions

plasma, expansion, a continuation of the outbreak of an evolutionary explosion, initiating as a dynamic foundation of being the attitudes of any individual, group, phenomenon, the structure of the matter of life in existence, gaining new lags in the fight against cancer and other factors negatively affecting the dynamics of $m <(m1)$, the rotation of material functions people interested in life, as well as creating the dynamics of a hyper instrumental, hyper industrial, hyper tool, hyper material, hyper phenomenal activity in initiating, supporting, creating/designing/building the foundations of life, as so far they have been more or less developed or transferred to a greater or lesser degree depending on the conditions (endless) environmental opportunities.

This is how it is further algorithming, improving, prompting steps literally and figuratively, adding, provoking, and proposing changes to the accelerator based on action, and not some mythical "solid" matter environment ...

God is in every action. Action as a spark, a miracle of creating an evolution of life, the attitude of real religion and science, and not a shallow translation of some existing structures that ... do not exist!

Yes. It is taking over nature's divine competencies, pushing the boundaries of cognition, action, and not the wall of matter.

Because matter is action, each further action breaks the following barriers, the shells of the shell (eggs too), which is in our ignorance, powerlessness, dislike, illusion ... breaking the illusion ... matter is an illusion, and the action is true. Let us try to apply this principle as a foundation and guarantee of results in any economic, scientific, or artistic initiative ... it is this kind of abstract artistry of matter and life. It is often an ambassador for hosting a dynamic, responsible, action-based, appropriate response to the action of the environment of matter and life ...

Divine natures, processes, passages, gradual

Not by fate, but some consequence, of reciprocity of the phenomena of the processes we enter

There is always some causality to which we can, and we add not only a symbolic 3 cents, and sometimes more, changing trends for better or worse.

This is the sketched picture of taking over the natural, divine factors.

Disenchanting the walls of nature's "geometric solids" ...

It was capturing more and more layers of phenomena structures, tooling processes, ... jointly instrumentalizing nature, that is, co-processing nature, through the infinite potential of actions as the unlimited layers of nature, God, matter, evolution, economy, science ... of life.

Like Mozart's music, it has managed to capture what could not be caught in the creation of music before.

The same applies to new grammar, the language of processing phenomena, structures, and their layouts. Capture the same with EskmDAA in the descriptions of action processes in relation to the material-primitive shallow description/explanation(or just lack of it) of the cause-and-effect "structure" of the course of events.

This is. It is pushing the boundaries of God, nature

Nature ... Pushing down irresponsibility, increasing the depth of interference ... ridiculing the level of participation in nature, God...

CHAPTER THIRTY-ONE

Further open sketches, rough notes about the non-accidental course of events, i.e., the possibility of reversing the given trends.

Further open sketches, rough notes about the non-accidental course of events, i.e., the possibility of reversing the given trends, if we systematically and ruthlessly apply them to work …

So it is a more materialistic, structural approach than a random, … nihilistic approach.

Accidental processes and the process approach to phenomena break the so-called randomness, which can be fully controlled by … structures of matter, because they were, are, and will be only a secondary effect of materialization, structuring, or rather the processing of the structure, depending on the activity, initiative of the interested parties, including our processes …

It is not about changing the environment for more or less immediate, particular goals; still about changing, influencing, constructing mechanisms, and hinterlands for the evolution of the environment, changing this environment directly or even more deeply. This is the proper thought, strategy, religion, science, and attitude towards the creative forces of the mechanisms of creation, the micro-macro-cosmic neo-creation of the whole matter of life ...

It's not about guessing mechanisms.

But directly building the tools of the base of changes, influencing the organization of directions, development forces of all factors, including artificial factors derived from interventions, processing (all factors/elements are

artificial/manufactured and not derived from "some natural sequence of things" as natural) – on the properties of matter and life, including light, cells, atoms ...

And it is also about encoding this new thinking over or rather a hypernatural, hyper evolutionary, hyper ecological, hyper economic, hyper scientific

In these thoughts and strategies, actions generate further social, economic, scientific, religious, political, individual, and collective activities on a full scale in small and large enterprises, in their genetics, code, pattern, and model of operation.

Yes. This is the artificial production of the thresholds for the development of evolution, the matter of life. Rather, the contribution to the existing more or fewer processes> structures> processes2> structures2

processes 2a> structures 2a, processes 3> structures 3, processes n> structures in single and multi-threaded discoveries / interpretations / applications, or rather processing at given stages of activity, back-up ...

Like an artificial "moon/structure, an artificial field, an excuse for exercise and the development of the evolutionary base."

Artificial moons, soups, evolutionary, specific technological, structural, production explosion as the following process contribution to further layers, atomic numbers, a cellular matter of life in selected directions ... improving this process without looking at fate from other "gravitational" forces

It is shifting the boundaries of evolutionary nihilism, i.e., not looking at the so-called inviolable laws of nature. It is the production of artificial thresholds, or rather a platform for further and more effective existential protection and development, and thus for additional levels of physical, biological, economic, social, and political existence ...

Civilization, economic, and scientific directly in the model, pattern, patterns, and assumptions in the boundary views of the existing lumps of environmental matter, the boundary views of which are the derivative of activities, processes, and further shifting of the boundaries of their influence, impact, a transformation of further deeper layers of the environment of the matter of life...

Society and its group activity express our identity and boundary in structures by processing the derivative of materiality as a further basis of EvmDNA <Eskm1DAA (m1 = Evmdna), which is in the further of this controlled evolution, an easy developmental threshold, but already from the position that Eskm1DA1A = EvmDNA1A = EvmDNA1A which in further processing will be the basis for the next Eskm2DAA, and so on in this neo-ecological, neo-biological, neo-physical approach m, E, DNA, c, ...

dream / action procesing neo / matter

Dream also determinate neoevolution that sparks further

involved, influence, size of the further effect of the matter, and dream from nothing ...

We are talking about the further continuation of tasks as a determinant of neo/materialization.

Not neoevolution, but neo-processing of structures, but the opposite in the construction of the existing matter ... the re-constitution of this matter

Statutation, neo statutaion of processes

No longer affirming "absolute truths ...

To exceed the nature of the cosmos of matter and life, we introduce and implement our own structures, properties, conditions and laws on a full-scale.

New mathematical map/ graph structuring/= processing -stuffing the structure with the process.

New mathematical map/graph structuring/= processing -stuffing the structure with the process – mathematics in breaking down every resistance of the phenomenon and structure/system ...

We continue with the further complex, changeable, not always transparent, but always an open description of a specific divisible influence, initiating changes effectively dependent on the external/ internal background, artificially or natural – natural, i.e., unchanging/predictable – influencing the opportunities, the strength of flexibility of the literal materialization of effects for the evolution of process-complementary

conditions to date ..... carriers of given areas, space, matter, and cosmic micro-macro life. In this case, we are talking about mega-economic properties = neo-evolutionary influence on infinite structures and their endless potentials hidden in unused/new processing options – a kind of subcosmic map / multi-cosmic micro-macro...

A map of a specific domino effect that can in each case be appropriately applied to change the world forever in the possibly productive / band/mass – ... chain reaction – that is, the correct road map in a neo-evolutionary way.

The endless processing of internal and external connections, thus changing its structure and properties in a possibly productive, ... IT way.

We need a machine, an autonomous system, a permanent accelerator supporting processes at the macro and micro level – the whole industry, economy – their parallel sector, bypass scaffolding, funding platform, supplying

materials, or more precisely, continuous actions/processes supporting and developing.

It is not about abandoning hobbies ... strenuous work but rather the mentality of perceiving existing structures and phenomena. Or even rather, to relieve ourselves of the monotonous activities of sustaining the networks of evolution on the surface of the cosmic ocean.

It is about creating a production, organizational / IT megamechanism, which is to gradually – as it is – but faster and more efficiently – take over in initiatives in the trends of the matter (and) of life ...

It should not be idle but accelerating and supporting indefinitely.

Our task is only to apply this evolutionary tinder in the correct direction, use it more and more effectively, and direct the same driving tool system of the hyper global, total, hyper cosmic civilization – ... it is us, not the cosmos, that is to determine the laws/=rights! to the matter of life ...

The more precise the targeting ... the process set, the so-called structural effect more beneficial for the target person,

the better facilities in this direction.

Randomness itself, statistically randomness does not increase the chances structurally unless in the sphere of optimization – acceleration of our own plot – three-field ... – the material background of the new sub-cosmic

mathematical/statistical/geometrical presentation, steering, "informitizing" opportunities and directions – not only passive mathematical interpretation (peculiar religious mathematical justification of one's – illusion – reality) but its hyper adequate application –

the subcosmic supercosmic mathematical approach to the opportunities and directions of tuning, setting directions/trends/ principles accelerates the development in every area of life of matter – from the so-called internal always infinite process potentials for/from the structure – or rather structure from the process eludes our standard expectations, observations ... cosmic, material, biological (atomic, cellular, cellular functions, atomic functions hidden under the structural plane, and in fact in processes with infinite potentials in

attitude to expected or unexpected calculations, attitudes ...

Models, charts of acceleration or deceleration control, refinement, selections of processes, directions, and background of the cosmic suspension, over/under cosmic / micro-macro cosmic

sub-material / supermaterial

.....

the strategy of calculating the process room, creating new matter, sub-matter, and matter underneath the broadly understood Mendeleev table

in the broad sense – above the classification m, i.e., m1, m ... according to EvmDNA <Eskm1DAA,

above the DNA of matter, i.e., free process modification – anyway about the process basis of those already classified too -DNA <DAA

Further, these new processes structure their new – hidden potentials/properties, properties – literally – taking over their functions, laws/= (worked out-responsible not passive action)rights as already their own supernatural

(because natural – that is, not entirely known), ... and already classified/ranked, supernatural known in our area, the potential of giving it a direction of expansion beyond the natural/hypernatural scope of operation – using,

by shaping and changing them infinitely, and what further changes these properties passively / actively.

... Further, decent, ... economic, political, ecological, religious switching on (switching) in (not) existing model, geometric, mathematical – clarification, and actually also shaking the tissue of matter/life.

Not the so-called fate, but the neoevolutionary racing, the initiative, and directions of activities influence directly into the further shape of the cosmos of matter and life.

Previous technology, – and natural = "not created -"

Mathematical geometric rearranging, TLE technology, or direct infrastructure facilities, where, for example

The artificial shifts of the asteroid trajectory indicate – an example – always shows the infinite possibilities of processing the environment of structures/= phenomena (always based on the endless potential for the processing of every structure) external and internal inner and internal based on the current system of phenomena and structures with which the effect always occurs (evolutionary) Domino creation. The issue of creating ... Mathematics geometry modeling, processing new classifications, not only on-time statistics ...

Continuation in mathematical/ geometric – interdisciplinary plan/ map of space processing, cosmic control.

CHAPTER THIRTY-FOUR

Circle in/on a circle – a model of chain dependencies and an infinite potential in influencing and processing any structure of matter and life.

Streaming circles means a model circle EvmDNA in m1 in the Eskm1DAA circle, i.e., hyper-stream("hyper string") interpretation of the processing of matter structures on infinite dependent levels of physical, biological matter ... specific overlapping civilizations/interests/evolutions on infinite levels and not physically transparent behavior ...

This very extensive and challenging chapter of the sketch has to be tempted to pioneer the indication of the mechanisms ... of movement of the world in a relative interdisciplinary perspective ... application.

A model of participation in the infinite open circulation of matter on mutual more or less interdependent levels, employing streaming processing in controlling all processes and further structures of matter and life, and further as a new language, grammar, mathematics of defining, establishing relations in any public and private enterprise ...

Model, mathematics/geometry for conquering matter (of) life of the entire cosmos

for ever higher levels, all life and matter factors, life matter, micro / macro matter of the cosmos.

... Mathematics for engineering means the production of matter -life of the micro-macro cosmos, means – mega-production, mathematics of evolution, i.e., a tool for controlling, managing, automating, and autonomy of the background of an increasingly deeper infrastructure of the environment of matter and life, the properties of matter and life, and that is all process-based; the mathematics of processing dynamic – not some solid -structure, external and internal processes are determinants of all matter of life ...

Our mathematics as a building block

where the existing mathematical models are as if halfway between the so-called real and infinite potential process structure

; Our conceptions of strategy ... production for literally the actual correct trend, the determinants directing the development/ evolution of the cosmos of matter and life, the matter of life.

Maybe it is not about immortality, which is the goal, or conquering the entire cosmos, controlling its transformation processes, which is also the goal, but simply imposing, accelerating trends without unnecessarily burdening the load, with less and less burden for us, and a more significant burden for the opposite phenomena, with their incorporation, or exploitation, neo production, = neo evolution ...

The point is not to legitimize the state of nature, to legitimize nature, but quite the opposite

Nevertheless, it is about taking over its functions ... not involuntary further mental submission ... justifying some system of forces, structures, trends ...

It is about working out the process-oriented structure of a given structure as the basis for further activities, opportunities, and potentials – infinite potentials.

; that is, about further shifting the limits of intervention, responsibility, and law, automatically shifting all the laws of the structure beyond the limits of the current system. Any so-called holding back in this process removes its position in the always dynamic (always process-based, although it is often not visible ...) structural environment.

Multi-world / multilevel interpretation by processing any structures and trends depending on the level of processing intervention in the micro-macro structures of the cosmos of matter and life.

Processing interpretation of ... dark matter, i.e., the matter of factors, unknown, invisible structures

affecting gravity, etc., as if from a different/unknown circle, level of dependency

Interpretation of the material processing of structures ... not just atoms ... – the world of the electron, but rather its universe, where its streaming and other structures can be seen only after enlarging it to the size of the entire universe known to us, perceptible,

However, the dark matter that is imperceptible, not created, or to be created by our or another civilization/organization/system, i.e., contributions more or less, or relatively less transparent in controlling processes in accordance with or in defiance of any laws or logic.

At a given time, structures .. processes, intervention, interpretation, and intervention in the structures of matter.

Unknown particles .. = dark matter ...> gravity ... which need infrastructure process support

seeing, shaping further, deeper structures .. processes> structure> processes> structure ...

time, space, and processing –

it is they who determine, not the so-called structures ... Structure versus processing ... that is, a string/process continuity that interprets the infinite potential of trends and controllability of the mastery and life of the entire cosmos.

To be continued in the geometrical control, processing of processes, structures on any / infinite level of intervention, and the dynamics of this intervention.

CHAPTER THIRTY-FIVE

Any contribution of matter/life and other issues has a basis in purposefulness, density, and power of processes.

Any contribution of matter/life and other issues has a basis in purposefulness, density, and power of processes that lead to revival to the level of properties depending on it.

It is the revival of matter because there would be no life without it. Pointing to the methodology of taking over its functions, simply working with an expanded infrastructure base, which is to automatically take over the structures of the world micro-macro cosmos of metrics and life – but of life of matter, matter of life issue of life, issue of matter by life by action – by the process by more competent string "melodies" – a new DNA code <DAA for the life of matter by the process

....- re / naming the matter will facilitate the concept of objection to this matter of facts

dynamic   (mathematical/geometric/economic/physical/biological) indication of the correct points, lines, rings, circles, overlapping flows, intersecting processes of matter, creation, circulation, continuation / = life of matter, for a change in our process activity – knitting from this the crosses from this flint of spark and fire of life of people, stars …, in the universe of galaxies, the inner universe of electrons, cells of "different civilizations/ interests! processes not always in line with our expectations", … (but our civilization may also have something to say!), with each individual and separately, not by force freezing it in any frame of the model, but in dynamic, responsible changes, with an infinite speed of the potential for change …

Not transparent, but / because of it potentially infinitely

it is

is it random – not invariably contributory, but always to create an infrastructural advantage in relation to the existing structural event, but permanently suspended, tied in-depth, and fully processed… proper dynamics control of dynamics from an infinite fan of processes (dynamics of infrastructure/organization/ flexibility, their strength)

quiet employment, applying mathematics, grammar but also grammar feedback … the law … contributions to process infrastructure –

there is always trade in this process, i.e., costs and profits, but no longer a structural wall, but a soft, infinite maneuver potential that moves further the construction of the so-called ancient, not structurally active

contributions of the past to the further activity of the process hyper-structurally (beyond physical laws ..)

it is about systematic, not random, actual presentation/ rearrangement of the world; in short, and in the longer perspective – action, setting the direction of processes/structure processing, it might not seem too high a wall to jump over / penetrate (literally), not looking from a linear perspective, a contributory historical perspective, but an infinite potential for the coordination of arrangements and initiatives simply the mathematics of dynamic layout of maps, environmental plans with infinite potential, and also with our infinite potential to change it.

Mathematics has no disciplinary/mental boundaries – illusions and actions – it is ruthless

the only question is an application, production now, later, according to which paths/streams (strings) of process development disrupt the existing systems – contrary to our interests, building the successive carriers of our material, life – literally and figuratively – interests.

Process not/= structure – as if not static/computable, possible, impossible, constantly refuting values ... creating others in an indefinite degree and not relative! ... but for some, it will be like observing a geocentric system – (seemingly) retrograde movement of Mars in the sky ...

Structure past <process future <future structure (any matter/ issue) <further process <and our further very developed processes/ structures issues of life and matter...

process and structure factors from different issues in the same or different direction

not always transparent but always open deeper further stages of intervention, observation of micromacrocosmos of ... micro macroeconomy ...

the transition zone from structure to process, from this process to structure, and so on in an increasingly efficient, lightly controlled autonomic procedure (as you usually do with your own movements of the hands, legs, and eyes ...), less and less overloading the mental decision-making system) enabling more efficient decisions, thinking, religion, faith, science, economy, politics ... un/ transparency of right action, for changing properties of matter, means matter by action, process as determinant, the proper core of any effectiveness or it structure effectiveness model, ... of matter and life ... means Esk future always> Ev past although future Esk can be already of past Ev in future that means need one steady action for developing effective economy – efficiency of/for producing from matter – DNA of matter ...

-the issue – the mathematics of the organization of dynamization (literally also! – exploding), the geometry of evolution/dynamics of matter and life

infinitive un/transparency DNA by past old structure but infinitive transparency of our DAA by the future process ... so changing properties / = matter by process ... means re / neo/ structuring of matter/life... thus the life survive, developing life mechanism

Infrastructural process facilities as an internal/external basis, i.e., a proper material layer at a given level/layer of interaction, that this visible/invisible layer – material properties ... this invisible process-related invisible is, in fact, a determinant (animating spirit),

this envelope, but infinite; potentially in the process infrastructure facilities

... development limits, however, are dependent on the level and dynamics of this processing.

To be continued with the chain background of processes for, in, of matter of ... his algorithmics – application and further implications.

CHAPTER THIRTY-SIX

Without responsibility supported by an entire, very effective operation, a specific energy-material flow, there is no right to development.

Without responsibility supported by an entire, very effective operation, a specific energy-material flow, there is no right to development as an basic attitude of existence – the algorithm of the function/fusion of matter (and) life...

Your life matters ...

or rather your life is matter of process...matter is processing of diffrent stages of perception and action

through creations and flow, infrastructural competencies.

Just as in the absence of political responsibility/competence / active action, one is politically dead, it is similar in physical, biological, material, evolutionary, ecological, and then in life, existential ..., and economic feedback ...

To what extent we will responsibly improve the environment. We will be saved to the extent, that is, beyond the existing barriers.

Prayer ... learning is just like try putting thought into practice – literally.

At the level of a given discipline, physics, biology, economics, which are to coordinate each other ...

As far as this responsibility is concerned, we have the legal foundations of the matter of .. life ...

Responsible and constant building based on actions rather than willingness increases the depths of matter and life.

Only the hypernatural, hyperphysical, hyper economic, hyper evolutionary attitudes give rise to the creation of

the evolution of existence.

Here, a crucial factor directly increases the effect

of an application of the direct mathematical algorithm of development is an application of the scale effect in investments, operations, active support of responsibility is given by the actual scale of undertakings, the chances of the application of laws..based on the size of the scale of applications, which in fact will reflect the hyper-evolutionary responsibility – hyper natural – hyper economic – hyper ecological – hyper cosmic – hyper biological – hyper material ...

Responsibility is a flow; it is an active participation in the entire flow of processes influencing the creation of matter based on objective, constantly accelerated pro-development infrastructural, hyper-natural competencies, i.e., ahead of other more or less accurate natural models, – of natural barriers ...

flow is life ... – this is a continuation; in this case, I summarize more, set the tone, than give precise contours to the frame

Then continuation summarizing ... to act, not to stand in an entropic place / entropic black hole (and not anchoring to some absolute models on which we are to conform – forget this scholastic language forever – its grammar, its course is to be changed concerning the treatment of the environment, and the related responsibility as a determinant of the language of science, economy, politics, religion ..., us in this environment – this is the new Mendeleev's table ...) because it there are real models of illusions, but open topics/abstracts for further processing, implementation, coming out of mental, natural, physical black holes, coming out of model illusions, in new economic and scientific strategies, producing your own supernova of the old doom, technological explosion, organizational, algorithmic for the new Ev, i.e. new process foundations of Esk matter (and) of life ... So continues with further suggestions, pointing, drawing, and explaining these circular flows/strings for the creation that are the fusion and decomposition of the accumulating old combats, captured in the material effect, which is a component of these long-term fusions of natural and artificial emergence (with our participation / responsible/responsible)) the cosmos of matter and life ultimately on a scale of infinite size and speed ... that is why this cosmos is getting faster and faster, expanding with(out) our developmental line or other.

The question of participation, commitment, i.e., real active responsibility with every step, breath, and initiative for every part of the macro and microcosmic universe, in which we are a civilization in an economic, physical clash, coordination with other civilizations (hidden in the universes of each electron and other particles of matter and life, i.e., systems not always going along the way according to some model constant, but variable interests not always consistent with a given model – an illusion of development or entropy ...

Flow -> creation -> creation -> flow means (matter (of) life) fusion of EvmDNA1 + EvmDNA2 + EvmDNAn = EskmDAA, where further EskmDA can be a material component of mx further different branches, fusion n EskmDAA for higher, further, deeper levels interventions, accelerations in the physical, chemical, technological, economic, organizational, political departments, in the global fusion – division of matter – the matter of life based on this processing and its constant and more powerful infra / structural service.

Not passive ecology and not symptomatic passivity, but active, creative intervention, i.e., the increase of hyper-natural competencies for neo-materialization, that is, to stay in a dynamic world – there is no life without it. Hypernatural responsibility is life; otherwise, it is evolutionary and direct suicide.

Furthermore, creativity is related to the flow and feedback, i.e., the increase in our invasion competencies in our own process fusion between the old EvmDNA and the new EskmDAA.

Fusions, the construction of our own matter, the basics of linear matter on action, involuntary processes, i.e., natural and artificial processes, i.e., ours, our participation, i.e., support for our position, and the position of the environment, not in words, but in the fundamental assumption of responsibility for the material that will be taken over – flow-like, progressive – mergers, reconstructions from the process of old marteria, also coming from earlier processes of mergers and divisions to the subsequent postings through mergers, sharing it in a formula, EskmDAA model, a given level, a given topic/issue, to higher levels, where in a further level – now EskmDAA1, there will be a material basis in this process for further processing, the fusion of EskmDAA1, 2,3, n ...

Fusion/division from Ev to Es make matter, is matter..is the existence of any matter ... of life in any dimension.

To be continued with the artificial/hypernatural fusion (processing) of life and matter.

CHAPTER THIRTY-SEVEN

Algorithm for creating artificial niches in the transformation of matter of any phenomenon/process, its properties at the indicated level,direction.

The algorithm for reneostructuring the matter and its properties, including the so-called basic inviolable models of matter and life. Real science as a study of the acquisition of nature as a primary, existential paradigm for further activities and initiatives, as a literally legal, economic, physical, and mathematical basis.

Continuation of studies – not just reading – unless one has no idea, no material point ...

EvmDNA as m1 for Eskm1DAA, and Eskm1DAA

as m2 for Eskm2DAA,

However, creating in the whole meaning of the word, as the basis for development, the existence of matter (life!) m contained in EvmDNA, as further material m1 for the further development of Eskm1DAA, etc....!

Is it a fight against nature, a replacement for nature? Not! This is the essence of nature, the nature of the environment, ... evolution, ... ecology, as a responsible, active attitude in participation with full rights supported by actions and consequences in each case of the so-called static or activity, passivity or activity.

However, it is a frontal attack on the hitherto found foundations of the entropic scientific, economic, civilization, and individual model. It is not a simple rebellion but simply a possibly complex presentation of suggestions/studies indicating the structure

infrastructure – an artificial niche, human activity, and other systems (civilization, particles of nature) in every

field and space of matter, life, for life.

Studies that is, searching, creating one's own ways, one's own intention to achieve one's own goals, dreams – always unconditionally!

, at the same time achieving the so-called scientific, economic, and civilization goals, and above all for them, precisely those goals of implementation!

This is the material for studying, lecturing, and making for students, and workers -"robot/nik-" in Polish, literally material to study and process literally based on the material m / for the new material m1 = EvmDNA for Eskm1DAA .... activity and obstacles in the (not (/) realization of the EvmDNA <Esk mDAA process, but as material from the point of view of statistical, mathematical, economic, physical studies, experimentally for further modeling, modular m for Esk became further matter, further envelope

of economic and scientific, mathematical ...

We are talking about studying environmental matter processing.

Process, i.e., dynamic mathematical description -> process future, not present.

Imposing the structure / or, rather, processing, i.e., infinitely active life, not passive description of processing – the dynamics of activation description of structures, i.e., about infinite spatial dynamics to the inside, i.e., a kind of stuffing of matter! That is, not to any barriers, not entropic, but rather a catapult model of the universe's dynamics of matter, biology, physics, and chemistry.

Not closed models enclosed in any formula, but rather the lack of a formula, a non-transparent mathematical formula that seems to refute the old school of physics, ... and openness / based on technological/economic/organizational/mental changes

applications in the political role, grammar, religion, mathematics of matter, and life of the micro-macro cosmos

-> religion / ... science deeper understanding, its fuller interpretation, and application, argumentation, and definition

are processes and not structures (including our non-structures), .. grammar, the economics of the structure of the processes of values .. relations of the value of the quantity possibilities ...... of matter and life and not looking at "structures", "resources" but with the infinite developed potential of developing the foundations of the structure of matter and life.

This is the extended use of math, grammar, and geometry. Not limiting it, but artificial as a catapult of changes on the scale hyperevolutionary, hyper industrial, hyper neolithic -> infinity of space and time of change on demand – immediately ...

in an infinite set of overlapping circuits – an infinite stream – a perpetual motion machine – processing, structuring, generating infinite quantities and speeds.

The question of catching (... producing) in this process as an involuntary factor and cause of an effect. This is hyper evolution, hyper ecological, hyper-social, hyper-religious, hyper material/life

..The universe in every particle separately- infinitely potential against any laws and barriers – structural freedom of string streams -> civilization processes of other systems where the so-called classical laws of physics may not

be significant, e.g., in our case, the plane is heavier than air and does not fall, the same may apply to other aspects subject to feedback factors, not always directly subject to the so-called fundamental laws, observations, and experiences.

Like the structure of the matter of life for its processability, the processing of the environment from the inside and the outside of

this hyper-dynamic image of matter and life

wave its rights, its knowledge in this EvmDNA/EskmDAA matter ...

CHAPTER THIRTY-EIGHT

Any natural or artificial action directly corresponds (is directly responsible for) to the further metabolic process of any matter and its properties, which in turn influences the further effect of this action. Any process shapes (properties of) any matter.

Dynamic, i.e. a process approach to the formation of matter, its infra/structure, and then its properties, and then the further course of the process, i.e. processs> / = matter, means the process is matter.

One keeps looking, shaping the environment, determinates = affected by processes – what was the first egg or chicken? ... but the process determines the evolution of matter ...

In this dynamic approach to matter/life, one can distinguish the issues of finding negative, positive process factors that contribute to the structure (properties) of a given matter/issue to the extent that is absolutely dependent on this total set of these process factors.

Not only finding but applying the right paths, crossing paths appropriate for the total flow of positive and negative "energy" vectors.

It is even about the constant, systematic, automatic, sometimes decision-making setting of the system focused

on economics, the algorithm of shaping the co-shaping of matter/life environments according to our evolutionary aspirations ... not only other stakeholders of this matter and life market.

The participation of different parties in this process of realization of aspirations is a set of factors that contribute to the matter of micro and microcosmos environment.

Factors that make up the properties, the change of the trend in relation to and to the extent always infinite/total and constant...

So the collection of these factor flows, rings, and circles of life/matter may not always have to comply with the so-called established laws of physics, chemistry, and biology – a specific non-transparence due to civilization/separately trend process! hidden factors hidden in the universes of the micromarocosm of matter in each elementary particle separately!

These are civilization factors, that is, building systems not based on the classic observed trends of the so-called "natural" micromacrocosm.

Hidden generating factors, factors of flow, changes, evolutionary trends of a given level, and direction of a given issue of matter and life.

Decision factors not always, and even obligatorily, must be against entropic trends from our point of view, interpretations at the civilization level, personal, economic, political, local, cosmic ... and at the level of other infra/structures = cosmic micro-macro processes – processing preferences.

There is no generally efficient planning of the chain of overlapping material/procession rings...

overlapping matter -> process plan of correct finding oneself in the multifaceted processional labyrinth –

This planning as such, which does not exist, as it was and is with the approach to climate processes in a global and individual perspective – management of global process factors, including our initiated processes

-> because we do not see more, they should be based on the perception of strategic knowledge, imposing trends in selected niches – which we discover and develop.

Without specifying a plan, there is no backbone, no specific embryo/origin from anything on which to materialize literally and figuratively. Re/neo materialization of goals

the existing ones are more based on the fight against resistance and not profound bold lower changes of nature, such as ...

Deep actual intervention, that is, taking active responsibility for the place in the micro-macro cosmos

Process responsibility directly correlates with the materialization of changes in nature, directly corresponds with nature, and with matter.

Continuation, developed, flow through the growth of infra / structural competencies – material based on a processing grid – in an ever deeper and longer perspective of matter/issue processing;

transformation, further creation/evolution, possibly brutal, but also responsible for negative/positive effects – just their costs, and not pretending that there is no influence, that it is the fault of nature, god, etc. – God, nature is in us, i.e. responsibility; as much as god, nature, so much responsibility and guilt in us – Responsibility based on deeds, processes.

..because of the shape, consequences of planned actions/process technological algorithm .... material hyper evolutionary hyper biological hyper material to control neonatura, shaping the environment as if it replicates itself self-developing systems build infra/structure more and more efficient, more precise, wilder / more powerful than nature

organizational and environmental preparation – infra / structural, mental, ... comprehensive, economic

-> spontaneous flow -> ..

-> matter neo <> supporting matter/life through the process of hypernatural fusion EvmDNA -> m 1 m2 <Eskm2DAA...

CHAPTER THIRTY-NINE

We must constantly break through the barriers of matter to live. Moreover, life is a process, crossing material barriers expressed in the structures of infinite, hitherto process relations. Life is expressed in the so-called hyper natural, hyper evolutionary, hyper material intelligence ... this is the god of life, processing, a sequence of materializing processes, over which one must constantly gain an advantage in order to live in freedom beyond entropic, beyond the barrier in an infinitely increasing degree and power of prefabrication / the neofabrication of matter, i.e., the environment as the basis of matter, its properties based on this process of life of matter ...

An animal as something more than classically physical, more than classically material, initiating beyond the material, beyond the established frames. The animal as a germ supermaterial intelligence

It lives because without this freedom above the material understanding of relations and phenomena and properties, there is no life, the determinant of which is interactive intelligence against the brutal so-called physical laws ... a given level of material and physical dependence depending on the intellectual activity of this animal, this soul hidden in the infinite potential of continual processing, which is always a further factor of the transformation of the properties of matter, matter, any temporal structure, to an infinite degree; the question of the correct application of the mathematics of economics of the vector system of this process plant.

Leaving the cradle of humanity, the old cradle of matter processing. Exit from the orbit of thinking/submission and entropic action, under the evolutionary matter of the micro macrocosm, i.e., the process approach to environmental processes hidden in the matter –

The real soul/animal, civilization over inter material hidden, produced, discovered, felt at selected levels of activity, faith, and then actions, vectoring, processing, its teams, which make up the materialization of goals, stages of these intentions – stages more or less so-called natural, artificial.

Catching this increasingly tangible model, graph, dynamic geometry of the architecture of life processes, i.e., matter, i.e., in reality processes, i.e., matter at given stages of the place of events, reactions, counter-reactions, as the basis for seeing, creating/ reacting matter, ... life as the functions of the foundations of development, the level of development, the level of matter, the level of processes, levels of life, factors of its maintenance, developmental or degenerating factors, regenerating, neogenerating factors that we have had, and we will always have an influence on as an obvious, conscious share.

Consciousness expressed in this way expresses our level of intelligence beyond the natural, beyond evolution, neo-ecological, positive, or negative.

Yes. We are talking about a hyper-evolutionary animal – a soul of civilization, a specific intelligent system beyond the material / hyper-natural of a certain level to exceed a specific micro-macrocosmic velocity – neocreational; as if by no means supplementing the existing systems of matter ... that is, process-wise.

Hypernatural Intelligence – The starting point beyond the entropic approach to phenomena,

Exiting a given level of the cosmic velocity range ...

A civilization animal is hidden in the so-called non-normative processes that present the matter of a given level, depending on the discipline, direction, strength, and level of a given procedure, more or less mixed, including

physics, biology ...

So it is about the economics of trading with this well-known nature, not the imposition of standards.

It will always impose them; it is about the awareness of the level of imposing mechanisms on interdisciplinary interinfrastructure trade with the structures of nature, the civilization of the peculiarities of nature.

It is such a dynamic, open, infinite architecture of the life matter – the life ('process) of matter.

The technological thread only appears at a certain level of process accumulation; it appears, is created, and evolves. Material-an assembly of matter to the next stage of this often overlapping process

the question of the proper participation of economics, hyperecology / hyper environmental perspective, or rather a hyper material/process approach to the environment – hyper physical ... hyper medical ... its treatment (environmental medicine – treating it as an organism for joint creation, evolution ..... not the sin of omission, not shifting duties and rights! to the so-called god and nature, which is also in us, and is also our part, on our part), but on the other hand, a classic medical approach to life, material-process! life carriers.

Organic and technological architecture – factories processing matter -> for our matter -> more comprehensive matter of a given further process in a mega-productive, total hyper-environmental degree.

It is only about a hyper-above-normative / above-natural / above-material approach to the subject, which gives a chance for real participation in a better procedure of the environment at the fundamental level of responsibility.

Continued in the economics/technology of the creation of matter and life.

CHAPTER FORTY

Capturing the matter in the project. Materialization of the project, the effectiveness of which depends on .... dematerialization...

Capturing the matter in the project. Materialization of the project, the effectiveness of which depends on .... dematerialization through more processual approaches to the causes and effects of our lives, thus materialized ...

We continue to study/ we design the subsequent contributions of matter/life of any natural and unnatural phenomena as an external observable shell of the reaction of intersections of processes, which are further derivatives of this material effect of intersections of intermediate processes, and which we can, as it has been and still happen to configure in obvious observation and faith, that it will give something... for the infinite potential of change, and therefore continuation, and therefore matter, and consequently economic, social and personal, biological and spiritual life in a more responsible context.

It's not about naming, illusions, but the intensification of process activities .... about servicing with the help of the so-called external systems amortizing the costs of a given material status quo/ intersection of processes, services, ... interests, as the objective basis, factual basis and a chance to maintain, develop one's status in any human activity, including and for life...

Projects over nature are a natural and obvious order of things in every developmental, existential, and even vegetative human activity, seemingly bringing no changes.

But the more efficiently they go beyond natural barriers, creating their own more familiar natural foundations, their own rules of the game, anticipating, inspiring, the more significant these changes beyond normative negative or positive can be.

We are talking about our visions and actions, even the least related to the so-called development. Any activity is a fertilizer, a foundation for the next ones; it is a change of the environment for someone's benefit, a loss that can be felt more or less positively on the prospects of the impact of existence, development, or degradation,

whether we like it or not... and it will always be at least with the most selfish, primitive point of view and action or inaction.

The chances increase when, of course, we take a look, we consider what has given and will continue to give a significant advantage over the current course of things.

Circulation, chaining, layering, crossing, colliding of the cogwheels of transformations, which continue to intersect with other wheels of the evolution of matter and life, causing constant material, energy, life changes, their further existence, development or degradation ... which we can also appropriately thus more or less directly shaped... more or less passively by these and other factors in this matter (and) life markets.

The project of the proposed changes and initiatives is a project, an idea to be implemented, which can be more or less global in defining, and shaping the environmental factors of the matter that we deal with, that we use, and for which we will pay more attention to efficiency, comfort, our realizations less or more earthly goals, with a mediocre or revolutionary aspect, which will increase the chances of the present, economic and sustainable existence.

A project, by definition, consistently exceeding the resistance parameters of nature is a project; it has the right to be called a project.

This is the efficiency and sufficient impetus necessary in every action, activity, phenomenon, and process field.

It's about the extent of reproduction.

The awareness of this implementation of the materialization, or rather the processing of this.

The greater the precision and the relationship with other factors and processes, the greater the material value, expressing the effectiveness of the introduction.

Writing projects, studying and developing virtual efficiency from goals, plans to matter, i.e., materialization, its effectiveness depends on this project in a given topic of matter connector in project processing ... which is contradicted by ...

... the sin of omission in fighting evil, improving the world as the main thread in philosophy, the strategy of continuation or non-continuation of hyper/evolution – creation or EskmDAA>EvmDNA or existence, the basis of which is responsible creation

It is possible to forgive this existential sin by oneself by building up, improving the hyper/evolutionary infrastructure/structure, and thus not sinning, i.e., ignoring these project hyperevolutionary activities in any field that will show the right responsible approach to one's fate and one's environment in the near future a longer perspective, on which directly depends the real perspective of the evolution of matter and, consequently, life, based reflexively and unconditionally on these processes.

To be continued with the outlining the dynamic architecture of matter, or rather the mechanism effectively processes it, expressed in any more or less responsible projects, i.e., with potentially adverse, positive effects of transformations of always dynamic matter determined by actions/processes, including ours.

CHAPTER FORTY-ONE

determination affects

bigger smaller project

more universal

more precise and purposeful

(or indifferent=the game itself) in determining simply an increasingly efficient hyper-infrastructural system of the world, conscious, affecting the above project of predicting

in a democratic system... the lack of it is a sin of omission – in human, social, technological, economic, political, religious, physical, and biological terms.

Processing plant efficiency (... not abandonment) is material efficiency

over other material projects, puzzles depending on the issue of flow, jostling, and not only on their own design or more private –

reciprocity of non-cooperation competition hyper evolution

-conviction on their own skin, at their own cost, goals, application of the linking tool of the infrastructure/nature processing project – whether creating it

Building and constructing the following infrastructural platforms gives a chance for further conquests in setting trends in neutral nature, whose tendencies are determined by directions, vector sizes – m, m1, m1a, m1b, m1 ..., m2a .... vectors of factors from centers less or less more familiar to us as well as our more or less distant environment.

These centers/environments gain their strength and advantage by building up the following self-development platforms, generally driving the trend in establishing this "neutral natural (process)

lever of the system's edge at the physical and biological level.

We are talking here about technological/=economic infrastructures, which will be the basis for the following scientific progress, which at a later stage will further affect the development of centers, a kind of autonomous centers that are to grow/raise ... following levels of infrastructural platforms

in this war of competition or cooperation with trends that are often contradictory, mutually exclusive physical and biological relations, which the so-called specific entropic or expansionary balance should be directed

Parallel, next parallel, and next parallel of the scientific, economic, social, mental infrastructural front

Just the mental hidden in dreams, aspirations, plans, and thoughts that are the steering wheel in dreams

It just happened because we believed

This belief is the complex contribution of the DAA....in this pattern of process foundations of matter/structures/properties on any scale.

DAA for a new matter of new life, continuation, and development in process/matter

Esk(m1=EvmDNA)DAA > EvmDNA

Own process facilities of newer and newer generations on/over the corpses of old EskmDAA/EvmDNA evolution systems in every activity, including cell, atom ... company ... state, personal ...

Not always effective, but it is always guaranteed .... but in the end, as in football, at the end of attacks in the penalty area – it is supposed to succeed when building a DAA process advantage over the DNA system of matter – at a given moment, the system seems to be unmovable or excluding or entropic.

Infrastructural artificial solstices

Layering Esk over other systems, schemes, crossing them...

It is about the aforementioned new language, the communicator of the translation of dreams

a kind of direct synapse of the system in translating, imaging, and organizing economically – economic in fact – consisting in economizing all structures employing processing, which is the basis for intersecting restructuring, "material" foundations of development and structuring / politicizing, i.e., the science of .... the civilizational community of physical systems, biological... to a direct degree, not always, or rather never, adhering to the so-called physical constants, relative and absolute biological constants of Einstein, Newton, etc.

Structures/processes and/then properties, not the other way around!!!!

That is, none of the existing systems can withstand the so-called new trends, and the processing plant rematerializes the structures of the environment.

An adequate system, a mechanism for effectively translating dreams into realizations, despite the so-called obvious/current calculations contradicting them.

Like David with Goliath, like a man with the whole (micro)cosmos...

Here the mechanism of absolute, direct control of processes is expressed, influencing micro-macrocosmic processes, their directionality, strength, structurality, and leverage in constructing the neo-evolutionary infrastructure, the neo-creational cosmos of matter and life.

It's like directly writing ... programs of infinite processing potential – it's nothing else than science, the language of taking over the structures of the entire cosmos micro/macro to an absolute extent / not relativistic attitude!

To be continued from being an object to being a subject in existential, evolutionary relations, i.e., more evolutionary and creative initiative through proper elaboration and categorization between matter and process, i.e.:

Matter as an intersection of processes, as unknown to be worked out at a given level – matter ... metallurgy ...

and then infinite levels of physical, biological, and economical processing/creation.

CHAPTER FORTY-TWO

In the world of living micro-macro blocks – animals/civilizations, of which we are the heir, shareholder, and creator.

In the world of living micro-macro blocks – animals/civilizations, of which we are the heir, shareholder, and creator, i.e., further development of the matter processing plant and the materialization of the process.

Utterly contrary to any laws, rules, observations, or decisions that we consider to be under our responsibility, total responsibility including generosity, mercy, innovation, being living, divine means responsible above any matters and processes, simply a human, not a wild animal, human not entirely block of matter, but ....living block of matter. Being this shareholder, the initiator of the flow, above the current tendencies, inertia, and forces of nature of this more or less literally living world living... Lego bricks.

This is the supra-natural, super-evolutionary, and creative, responsible force that rules the world, whose effectiveness and generosity depends on this divinity, the spark of divinity hidden, in each of us, in each of the largest, most tiny pieces of a macro-macro particle of the cosmos of matter and life.

Yes, this faith, this certainty is a fundamental paradigm of any trends, phenomena, and their initiation, of which we can be shareholders – we already are – of the matter – more or less active, passive, or more or less responsible.

Just like other stakeholders/civilizations/systems (also not subject to the classical laws of physics, biology...)

hidden in dark matter, quarks, and other macro and micro cosmos of matter and biology of existence and development, possibly degradation... at our own request.

This faith in life as the basis of the decision is the unconditional germ of further processes. The question of working out these infrastructures so that they take the proper shape, direction,

and then strength

infrastructural supporting further processes, flow, tendencies, and ... laws – which have to be earned, not expected – and not vice versa "scientifically/scholastically"!

although we have such rights/laws established for ourselves that we deserve, we work out ...

The flow of faith of decisions beyond natural, anti-natural ... that is, anti-inertial, anti-entropic ...

And not only the desire to influence the directions, and then the properties of matter based on flow directions – but faith, desires based on infrastructural/=hyper natural investments.

... writing programs, divine = responsible decisions! i.e., internal, external faith -> decisions of rights against natural resistances/ opposition/ods, natural negative inertia, i.e., writing ... against nature, against its laws ... creating its own facilities, niches, environments, displacing, treating positively, infecting a positive environment for us entropic... and only resulting from it proper deserved rights=laws to life literally.

Creating an environment for oneself, including all the matters of processing, flow...trend values, is a planned rebellion against nature's status quo and its constant justification.

Proper classification, process view of the matter, and matter not achievable in the process .... so far

structures or intersections consisting solely of

processing, operations, and their flows – the shaping of this matter is absolute – although it is not always possible to describe the processes currently, but only superficially, i.e., materially, establishing some physical and biological axioms, and the "laws" based on them, which, however, do not match the phenomena we do not understand to the very end, never wholly known, e.g., dark matter, contradictory phenomena in the world of quarks – peculiar quark civilizations, animals not governed by any laws known to us, etc...

And processes that are not achievable classically strictly materially – that is, anachronistically.

....Imposing flowrings of specific tax drivers of natural processes...

ecological/hyper-economic nature in the physical, biological process to modify these influences of the flow of matter and life in applying a placebo – our interventions on the levels, trends, and territories of action.

...process niche

-lack of flow closes-ends the system of life matter

the need to maintain the development of the flow as the basis of any properties/values/participation of matter and/of life in the physical, biological, economic, technological, production, ecological, ... political, religious, social, and individual process, of each individual together and separately.

And all this from a different point of classification

-1 invisible matter hidden in processes,

-2 process unnoticed, hidden very deep micro deep or very far away (macro-micro cosmos in matter or seemingly visible but unnoticeable in a pattern ... in intersecting patterns and systems ... dark matter proper finding these boundaries, creating boundaries.

This processing of the structures of the matter is crucial in understanding the proper participation in the creation of

mechanisms of evolution, the creation of matter
-a materialization of processes
-b processing of matter.
More on process mass/matter inheritance in the next section.

CHAPTER FORTY-THREE

The race – that is, the life – of us/our system with nature – the systems/=procedures surrounding us from the inside and outside

brutal economic/productive exchange, intense, relentless, but always with chances, as long as this exchange, this process(/=life) goes on.

The race with the so-called external internal more or less known competitors, possibly allies, known, less or more available/own, i.e., natural, natural, our matter hidden in processes, flows -> economic exchange of production, technology exchange in which we must participate – in total we take this part as a passive mass or an active process that materializes this mass into something created, deserved, worked out as a structural mass for now, or constantly worked out as a live process positioning as a process, as a guarantee of our life, survival, existence sustaining these activities responsible development responsible supporting all the necessary activities affecting the desired existentially (anti-entropic ...) shape of the world's nature literally!

This is a modern model for winning your position / existential property to infinity on all fronts of our infrastructure, structures against clever structures, infrastructures of nature, systems of nature woven into our system, and vice versa.

It is also a race, literally a race with entropic innervations, of the expansion of mass systems, more or less process (inertial/=mass) micro-macro of the cosmos.

The ruthless, responsible economic/production inter-infrastructural transformational fight will help this race.

Robust/vital economic exchange, intense, ruthless, but always with chances, as long as this exchange, this process! (=/of life) will take place. This is the complete process -> material basis of all physical, chemical, biological, evolutionary, economic, technological, production, civilizational, political, social, and micro-macro phenomena of the cosmos for here, now, and forever.

Further decisions worked out by us, determination affects

smaller/bigger project

more universal, more decisive

purposefulness, indifference/limits

of inserting an increasingly efficient world system into the above assignment of taking over the decision-making system,

meet the technology of economic, political, religious, physical, biological

efficiency of the processing plant material efficiency on other structures of processing the phenomenon, depending

the question of the flow of these projects' reciprocal abandonment

cooperation, hyper economic/ecological competition

conviction on our own skin, at our own expense and goals

use of own connector, tools, and flow projects of infrastructure produced.....

everything has its material plasticity

but this art is based on past, present, and future processes, which are still based on any decisions, programs that ..... initiate creation by still .... after the process

these repetitive processes, material plasticity, but also provoked by the greed of materialization, where a person can be even the smallest, the weakest, but still potentially and in reality always significant ... a driver/determinant, an element of the mechanism ... a lever for processing a given level of matter, which is also is the effect, and in this further case the cause-> our task is to bind any dreams, even the most selfish, egotistical – against any laws of nature,

into this mechanism of process matter –

its algorithmization, technologization, and economization in the indicated assumptions classically environmental micro-macro cosmic physical, biological as well as social, political, economic

-> proper writing/entering it in the hyperdynamic EskmDAA > EvmDNA model in each field, as ... coefficients ... of responsibility and effects.

Yes, it is. It is an artificial imposition, accelerating this process beyond evolutionary through/against evolution.

It gave the achievement of goals beyond the natural, above trends and evolutionary forces, of which we ourselves can be determinants and drivers on an infinite scale – purposeful above the natural, above neutral,

supernatural, supra-economic controller of all processes – super-ecological, hyper-economic, neo-genetic, hyper-natural wizard of creating -> new the matter of the new world, the new environment, the new properties, the properties of the new nature at any level of matter and life, i.e., the process is matter is the generation of matter – it is needed. We have a robot, a chatbot (since November), and a man with decisions to control this cosmic AI mechanism.

It's about the organization, action, economic strategy initiative, i.e., economy, workshop, deserved takeover of nature ... the earth as subject, as God commanded us, i.e., responsibility in action, matter not the impossibility

productively imposing evolutionary breakthroughs in the right direction

productive, but also socially, politically, scientific/religious – support for science – "religion-> science is systematization, recognition, and control of this earth subjected to us by god/=responsibility."

This process should be self-improved ... so-called accelerated ... self-education to be the master of matter and life, algorithmized

.....

-> it must be .... the underlying technology behind the cause

demanding uncompromising evolution of the matter of life

To be continued in determining, defining with the control of matter and life, their classification basis as a reneoconstruction superstructure, hyper neo infrastructural macro-microcosmos of matter and life.

CHAPTER FORTY-FOUR

Attitude -> EvmDNA ->EskmDAA = the miracle of creation/ processing despite "material" barriers, which are only material for unconditional/independent initiatives/->processing/actions.

Perspective, gaze, perception, and attitude to the inherited mass of the environment, generate any further action affecting this environment, further generation of systems of change, development, or stagnation, i.e., physical, biological, ... economic degradation.

The inheritance mass, the material basis, indicates further changes in this matter, not its cherishing.

This attitude is like a miracle of nature, which further shapes the dynamics of changes in this matter of nature to a completely arbitrary degree.

At > EskmDAA > EvmDNA

So it's about working out this landscape, this mass/matter literally – the realization of your dreams, faith, and for that reason, visions of the nature of God and Santa Claus in activities that are to materialize their functions according to Eskm1DAA > EvmDNA where m1 = EvmDNA.

Flexibility, ruthlessness/nonrelatives of intercepting the phenomena of Esk>Ev matter – all the theories of relativity and non-relativity of physics and biology are once and for all drawn into fairy tales and myths, which, after all, have always been a way to search, and not stand still in terms of absolutely micro-macrocosmic.

It is simply necessary to expand and apply this to various events and intentions, this formula for the realization of dreams, goals that may seem unattainable but are still achievable always and everywhere for implementation, and to an unconditional extent, if it is properly worked out, developed – economically evolutionary.

Responsibility/faith/god-divine decisions and supporting activities are constantly developed infrastructural competencies of an infinite/infinitely scale, and earlier the laws that are established by the infrastructural need for production, the magician of production, i.e., evolution, nature, hypernatural, hyper evolution, i.e., the momentum of the competence of responsibility – divinity.

Attitude as the determinant of the creation processing of "matter" that is anyway product of earlier processing and structure(matter), structure, through processes, and so interchangeably infra/structure … determines the size of the impact to this level…DNA but on the more prominent causes and implementations and levels…

This is a kind of miracle of the creation of creation

dependent on the attitude to the procedural material foundations – from beyond the perceptively established dogmas of a given level of infrastructural/hyperevolutionary competence.

Attitude to faith

But from the infrastructure, from the science of engineering, economics

Attitude >belief>science>

science>faith

Taking over the entire environment and faith/god/ responsibility depends on the competence of science and infrastructure.

In the return process

EskmDAA/EvmDNA hyper generic/genetic

Reverse bias can be the result

of a given stage of DNA/DAA – Causes, effects at a given stage of this process "material" evolution of m("mass/matter")

Hidden in the processes hidden in the broadly understood quantum, i.e., not subject to classical mathematical models … matter at the quantum level …

At – Attitude is an extension of the model, the pattern that determines further results, directions, and chances in this cyclical processing of "matter, including those confirmed in quantum relationships, but also deeper beyond the physical, biological, beyond DNA, i.e., determining DAA ...

Faith, belief, and attitude is a paradigms-

but responsible faith -> not define barriers .. it is supposed to be a catapult for the development of existence, a cosmic explosion, not entropy.

Faith in God, in supernatural forces over "material" is faith in God, in supernatural (means magical? ->means above our perceptions and experiences, but never competencies, that need one to work out) powers, it is faith in ourselves, our responsibility/ competence for the world hidden in God...., his super/natural laws (super

subjective) by not adapting but neo-infrastructural faith in God in the supernaturality of the laws of nature, its hyper evolution -> further superstructure in accordance with the development of infrastructure/structure, perception.......

Faith is supposed to support, not overthrow

that development means that we are not imprisoned by one systematization or another.

Believe in the future, not in retrospect. Increasing the impact of shaping its proper strength

We expand the boundaries of rights/laws, barriers, or rather gates to open the way ... material procedural barriers/borders of procedural dynamics

the development of handling the processivity of "matter?" -but no longer matter as a solid lump-

but phenomena, not the discovery of processes, their hard wall of ignorance hidden in the set of perceptions of some matter/mass/ block,

is the basis for seeing and believing in the forces known and unknown to us, the processes of the "structure of matter" of nature it is the basis for observations, decisions, and actions ... and not some constancy, dependencies

but depends more on .. economic/s, i.e. responsible action and cooperation.

Quantum physics proves the instability and, at the same time, the rules/lawlessness of relations in the entire cosmic environment, especially in terms of the strong paradigm of phenomena in micro and macrocosmic terms.

To continue with overtaking competencies is a responsibility, i.e., the economy of processing of "matter"/processed matter.

The matter is a process; everything we do affects the shape and operation of the environment inside and outside us. We are responsible for this environment, and we can develop this responsibility, which has a divine character, to create, to regenerate.

So I wish you all the best in this rebirth, giving birth to a God in us that is stronger and stronger with responsibility supported by this process, working for a better matter/environment and spirit within us and beyond.

CHAPTER FORTY-FIVE

From a geocentric system to a heliocentric system allowing for a more quantifiable, more efficient view of the nature of phenomena processes.

From a geocentric system to a heliocentric system allowing for a more quantifiable, more efficient view of the nature of phenomena processes. From the material description system of the environment to the process one, overthrowing the existing ones as a breakthrough on the scale of the neo-evolution of civilization and the cosmos.

The Copernican, Newtonian and Einsteinian models imposed another magnetic process cushion on the previous models, dematerializing, de-demonizing the existing perceptions and technologies, economics, … religions and literature, and language…" and the word became flesh".

The model of the process approach to all structures of matter and life raises the chances of the human development level as the next milestone of evolution, and creation, which they were with Homo Erectus, the Neolithic revolution, and the industrial revolution ...

A clearer understanding of reincarnation, the "structure" of the atom, cell, black holes, evolutionary and economic processes, as if from an increasingly flexible, dynamic level of evolutionary, creative, civilizational, religious, moral, political, scientific ... mathematical, biological intelligence, astronautical, medical, educational.

The question is how far we will be; we will mature to it, but also to what extent this new system of thinking will allow us to develop into more responsible and efficient members of the not material but process environment cosmos in more profound and further areas of our foundations of existence and development.

Just like the Copernican theory

that not the earth but the sun determines the movement of planets, so now that not mass/matter but the process is the basis, and the frequent cyclical basis of all phenomena of matter and life ... and even if we do not see this process approach to given matter structures, it is a matter of that we have not matured to this technologically, economically, mentally. We must build up this process approach, or it is already hidden.

It is a new look at spacetime, full of going back/forward and moving through time and space in an increasingly controlled degree, among other things, in the fight against cancer... and death...

After the Neolithic, Copernican, and Einsteinian revolutions, it is time for a new revolution/course that will increase humanity in relation to the phenomena of the micro-macro cosmos to the extent that it is practically infinite in terms of time and space.

Yes Sir. It proposes revolutions on a scale of evolution and creation hitherto unknown.

It omits the current scientific, social, and religious trends, which were and are at the level, or somewhat below the level of evolutionary and mental, ecological submissiveness.

To be continued with the description and schemas, charts, patterns, and formulas of this breakthrough catapult...that matter is a process with enormous consequences for any area of life and matter.

CHAPTER FORTY-SIX

Application, understanding, technological, economic, and religious maturity to this process and not material approach to phenomena, in turn, thanks to this inverted perception of matter and life, will positively affect further technological, economic, social, religious, literature, pronunciation, science – medicine, astronautics, ecology, philosophy, political – without artificial material barriers like having something that does not exist, because it is a constant, changing image of processing of .... processing ..., as a new tool/material – a new carved stone, a new stick, an evolutionary process catapult, or rather a creative process for humanity, the cosmos, the dynamics of the open cyclicality of existence, development in which we can participate to an infinitely accelerating degree.

We are talking about a kind of process magnetic/air cushion

in recognition of all structures, states, models ... which is infinite in the distance and depth of exploration of phenomena, the unattainable finality of material modeling of these phenomena, which can be harnessed in process models with the support of all areas of existence and development of the environment and man in it

existing, developing, i.e., actively processing, and not succumbing to some perceptively demonic walls of matter – the illusion of matter phenomena based on overlapping processes, some sets of which, by definition, result in material recognition of these phenomena in selected models – e.g., often recognized as permanent – hard geocentric, heliocentric and other-centric models of a given level of perception and usefulness, attitude, ambition, enclosing oneself in one's previous scientific, engineering, medical, economic achievements...

It is a specific system of phenomena with a two-cycle dynamics – process -> matters -> process -> matter, but rather process a -> process b -> process c.... where transitions between them are sometimes visible in static material models – there are specific effects, gifts from "nowhere", which often deny the existing trends of this cyclical nature, but also the volatility of the process chain of reactions ... they are a kind of evolutionary fraud ... because they have a creative – process character.

To be continued with sometimes quick short steps/articles/ charts in the non-material description of phenomena, static material models DNA micro-macro of the cosmos "matter" and life/=processing...

CHAPTER FORTY-SEVEN

Not a demonic definition of nature. The matter is processing... it does not matter; it is always the result of processing.

Not a demonic definition of nature. The matter is processing... it does not matter; it is always the result of processing, i.e., processing cycles are a component of matter, which may further be its further form of transformations depending on the processing economy of a given process structure.

We unconditionally continue further steps, economic and scientific concepts contrary to nature, or rather against the static model of its factors, shareholders that we are fully entitled to full responsibility for its further dynamics of transformations, which are the basis of its and our existence, in whatever form we would wish, supported by work, the infrastructure economy of this complex of process cyclical structures, as the fundamental basis, the nature of the micro-macro cosmos within us and beyond.

Processing means economy of nature, the economy of nature ->

Perception DNA – an attempt at an algorithmic description of processes and taking it over for further interventions, and cooperation in the economic field of nature, because the specific economic interests of nature's process cycles affect their internal and inter-relationships with each other with groups of process cycles – and what further means that each process has its own not some constant natural neutral unit, a trend model, but a dynamic system dependent on constantly changing relations in the relations between the process cycles representing, materializing a given state – that is, matter-... living matter DNA economic activity with the environment to the outside and ... inside.

This view outlines it more flexibly, more effectively algorithmizes the problem of ... materialization/=processing of your goals – materialization of goals is life, it is a dynamic image of matter, this materialization ... – process algorithmization will equal or exceed the algorithmic initiation of continuation by us, redirection these processes in proper cooperation, competition with other cyclic systems of process-matter, biological, physical systems... their, our joint economization of nature's technology...

This process determines life

there is this fluid boundary between process and matter, its further course...

it is the economics of space, the environment, economics of processes/operations, ... the economics of the matter. The economics of the matter/issue of life is life; it is

moving borders, inserting our own equations, your own algorithms of transformation processing cycles processing, which contribute to the materialization ... our goals literally and figuratively, which are the further basis for processing so that this matter does not die, ... materialization / or rather the procession of life – showing his procedural, fully flexible, potentially infinite process, the processing of life.

Algortimization, i.e., the influence of this macro and micro DNA of the micro-macro cosmos of the matter of life, hidden in the matter of processes, processes of matter, in models, and rules that allow for systematic participation in shaping the infrastructure of the environment, its algorithmization, it's DNAase -> DAAzation.

So, in support of science, a total environmental economy can come through massive E ecology, through really

gigantic E of the evolution of the cosmic environment with a really total purpose and character, where we are to become a really more active locomotive in its shaping – and not just a wagon, co-responsibility for its further microprocessing and thus further macrocosmic processing.

The rest is just subjection ... giving virgins to dragons, hearts to gods, to be eaten, to appease some simulated demonic ruler, the structure of the cosmic matter/life process ...

It is only about a worthy human, divine child, or simply the result of economization of the process structure/foundation of the matter of life, a suitable place in the cosmos, not as a fertilizer, but above all, an active initiator of further guarantees of development, and thus additional guarantees/deposits of the foundations of our, our environment of existence.

Matura economics is to open the gate and give the suitable algorithm; it is the right holy grail of matter and life.

Of course, science can replace economic mechanisms in managing nature directly or indirectly in materialization, i.e., proper determination of processes and process cycles that will allow for the adequate transformation of matter, the basis of which is this process economy – economic life – life.

i.e., DNAizaton, or rather DAAzation, economization of processes, further setting, control for more and more powerful distances and time

i.e., DAAization

e.g....... Trade with cancer by bribing him with something else to eat...

To continue in this non-scholastic not matching nature but surpassing it on all possible and impossible levels, including beyond the quark, cellular, galactic, ... light ...

CHAPTER FORTY-EIGHT

To be stronger than nature, than God, i.e. (macroeconomic) taking over responsibility, competence by deed.

To be stronger than nature, than God, i.e. (macroeconomic) taking over responsibility, competence by deed, without looking back, justifying oneself with some dogmatic forces and models of the world.

Multiple overlapping EskmDAA > EvmDNA triangles, which maybe not be a coincidence but an initiative, unconditional presentation of conditions, can be a significant factor in reviving these supposedly material but still process triangles, including the E=mc2 triangle...

E=mc2 and other "real/natural" states, as well as models of biology and matter in a more or less economical process approach of overlapping structures in the infinite network EvmDNAn<EskmDAAn as... just m...point, one of the endless factors m, in... or instead as EvmDNA as m1 in Eskm1DAA – possibly from, over or in EvmDNA.

And then there is absolute and potential infinity to choose the direction of expansion beyond the scientific, beyond the religious, beyond the mental/dialectical/literary framework, beyond the classical framework of the Nobel Prizes, Einstein's achievements, divine achievements of any imagination and potential.

Is it anti-science, anti-religion, hyper-economy, hyper-ecology, hyper-evolution, or the opposite?

Or maybe the opposite...

Despite any obviousness ... an independent presentation, a gift, or a wish as the determinant of all actions, preparations, perceptions ...

Or maybe it's just about ordinary macroeconomics, micro-macro of the cosmos, physics, biology, etc., i.e., full active responsibility

= economy for space science,... being God...

Evolutionary, mental gravity pushes our thinking to the trend of the so-called lower subordinate in our position, and in reality, a higher current at our perceptual, technological <=> economic level (mutually determining production/process technology and economics) – a higher level of "matter" of the cosmos micro-macro of the cosmos,

technological progress, rebellion, presentation, self-presentation, self-presentation, ...

(new) economic, mathematical, .., linguistic, literary, political, and religious perception kills, weakens, and destroys the current gravitational currents of the existing system of assumptions of entropic combinations of matter and life

crossing the barriers of classical science, religions of the end of the influence of subordination...responsibility -> in the factors of hypergravity responsibility.

Entering the DNA processing level of the process infrastructure, as if at a lower infrastructural level, to modify it and reprocess it, which will be the active build-up of DAA, or rather EskmDAA>EvmDNA, which has a dynamic complementary process character in these seemingly static relations DNA->body.

It's about freer, more advanced, more efficient control intervention – EskmDAA>EvmDNA infrastructural process conditions for processing at a given desired level of life matter.

The external DNA of the "matter" or DAA that dynamizes.

Dynamizing the infrastructure of the external/internal natural or artificial environment through DNA... process character of changes...

Attempts from the simplest algorithmization Ev<Esk

A given process of "matter" even in the slightest way intersecting its operational changes this process, its application aimed at direct or indirect acquisition, but increasingly closer to this model's goal of Ev ... in the shape of Ev, and we Esk will amortize this impact, increase it further finding points of attachment, in a constantly more and more ...algorithmized triangle according to EskEv – in search of overlapping triangular structures EvmDNA and our triangle Eskmdaa.

From the side of a direct internal influence on the DNA of a given process hidden in the "matter" of the problem or from the outside affecting the external infrastructure/casing body, mass, space, time ... m of this DNA.

Model of inverted two triangle complexes on one side Esk on the other Ev in the direction of transformations or a more complete picture of m a given level of processes in the external image/illustration of this process, i.e., m surrounded by triangles EskmDAA / EvmDNA ...., but also from the inside and outside of the DNA /DAA outlined in my second article this year.

Not barriers, dogmas, walls, some kind of mass, DNA, ..., emc2, etc. they are only temporary, operational, technological – but microeconomic – but not ultimately defining nature, God – but macroeconomics of taking over divine competence, nature

Responsibility/=competence pushes natural, divine boundaries from shifting fate and blame onto nature or God.

Enough of throwing responsibility on god/nature because it is depriving yourself of competencies and, at the same time, rights/laws literally and figuratively, not limited to the specific status quo

Emc2 ... DNA, which is to be only m1 in further Eskm1DAA taking over areas of cosmic power with the help of ordinary. .. macroeconomics micro-macro cosmos of matter and life, the basis of the matter of life.

These are further areas of potential for the development of civilizational, technological, economic, high-tech, and scientific infrastructure, which in this way, like Ford investing in cars/people, and further for itself ..., will drive the development boom, and at the same time the existential platform to a degree infinite, not even a million times, but many times more and faster, and at the same time safer in existential terms...

This requires a new attitude, and this is not about total expansion itself, but ultimately the protection of values, survival instinct, protection of values in the infinite potential of threats, which is to be developed on the endless potential of our environment from the inside and outside in protection, development, breathing.. us forever...

Not the scholasticism of nature – not catching up with nature, its repair in statics, but it is taking over in the process – in the post-process approach

creating a new environment – supposedly anti-ecological, so anti-evolutionary – it is a hyper evolution of an infinite set of cycles / these overlapping triangles of the process of matter – and not the materialization of the process! – their digestion, processing, in which we can more or less countably / predictably – perceptually, diligently work on this DNA – i.e., neo algorithmization of technology of ..... "matter" processes – place

...nature is supposed to be an addition to our environment and not the other way around – but it's generally a matter of trading with nature – of course, you need to work out something, organize to trade it effectively, cooperate ... maybe sometimes, cleverly win tricks with matter/mass or rather process " wall" of the infinite network Eskm1DAAn > EvmDNA=m1n

Competing with nature, as well as with a set of systems in us and beyond, their infrastructure not always so model, not always so cyclical with their above-trend interests and .. trends.

Continuation of the so-called process magnetic cushions/ processing of matter – of the so-called unbeatable states/models... Is it a rebellion against nature, against God, against the current absolute course of the history of things ... yes. This is called politics, that is, science for us – full-fledged citizens of the cosmos in every sense of the word.

CHAPTER FORTY-NINE

Pentagon by (-)Eskm1DAAn < EvmDNA = m1 < (+)Eskm1DAA triangle system.

From inductive,/= vain/ stagnation=degrading passive activities (economic, scientific ventures) to explosions of deductive initiatives and ventures imposing civilizational, evolutionary trends, i.e., artificial algorithmization/control of the micro-macrocosm means

(-)Eskm1DAAn <EvmDNA=m1< (+) Eskm1DAAn means process rematerialization, means matter means process!

Further, we aim at the layered systematization of controlling phenomena, materializations of processes that interest us, or rather processing of the matter that interests us, which is a layered set of processes/operations with a more or less visible wall/barrier ... hidden deeper/more profound, further potential. Useful in activities, projects changing, innovating the rules of the game -real game changer of our and other universes of life and matter, science and economy- from defense and support and/by expansion –

spends more on armaments research than on civilian research, wrong? If we are talking about defense, securing foundations, potentials needed for effective, safe existence, development, and not material and biological subordination, then the need for this pentagon of armaments/armament research – hyperevolutionary armament infrastructure needed for the effective triangular system of matter and life evolution according to our response attitude based, of course, on actions ... that is, the Pentagon strategy for explosion triangles of development and protection of life and matter.

That is, it is about increasing attention, strategy, research efforts, and initiative in total relation to the systems of

nature within us and beyond. We need this pentagon of strategy for scientific and economic ventures for a real/genuine leap to the following stages of existence, development, expansion, expansion, further chances of our existence ... on the (macro / deductive) economic, (macro / deductive) scientific level – small and large private/public scientific and economic ventures also need this strategy of growing a defensive infrastructure of the market for development and expansion...of /above nature...not just perseverance...

Internal and external infrastructural protective barrier

process hyper material made of Eskm1DAA internal and external infrastructural process pads that change the properties of a given state of interest to us, creating the -it doesn't matter ... any more... system, which is supposed to function on specific infinite levels, barriers of process pads on which it consists given "matter".

Economy/costs – economy efficiency Ev/Esk determines for us m DNA/ has a positive impact on the costs

we must in these two-step development steps – we have full open possibilities in achieving our goals – the issue of organization of this progressive two-step cycle – the accumulation of cycles/ processes at the level: cellular, quantum, molecular DNA segments, etc.

of course, the so-called vanity activities and initiatives can have a negative progressive nature – e.g., climate.

....Soul ->processing as the basis for further images (matter...barriers, problems, and potential)->process teams (matter...) -processing structures-specific image materialization of this soul/initiative (because matter anyway there is none – only perceived assemblies of process cycles at their respective level of reception and operation ...

spirit (without this visible, clearly departmental picture of the structure of processes..... warping of space...

transient systems-> indicating environment change Esk=Ev

... the balanced level of the course of the time-space particles of the processes .... and the geocentric system of thinking, action..

Process "air/magnetic" cushions at various levels of perception, systematization of matter perception, and activities in this area

that is felt/perceived or produced in terms of the classical structure -> but already under the processing system and not the material one....... moving on to further, or rather working out to further perceptions – actions.

The outline of the triangular cycles shows the founding, development, merits of other systems, development/degradation perspectives, and yet still existential ... and no longer exploration, but an explosion above the event horizon literally in any dimensions and imagination ... in planning scientific, economic, social, political, religious, personal...

This picture or the mechanism shown

potential for a new strategy towards any activities of processes from their insides, as well as external potentials transforming, processing a broader spectrum of rematerialization, i.e., reprocessing a given specific state of the phenomenon of the process of matter/=issue us interesting in any directions of our activity from artistic, economic, scientific, religious, personal, political, etc

Exploitation, engineering, ... production in the world of microparticles..., as if internal guiding of the industrial revolution ... parts of quanta, parts of cells, exchange economization, production of micro space environment of a given object, every object and phenomenon, its progressive de-evolution, neo-evolution control ... such an element of the element of electrons and others, quarks, cells, parts of cells, molecular parts of DNA, etc.

This triangle, the Eskm1DAA>m1=EvmDNA triangles, presents an infinite potential, a mechanism of mutual influence, working out the environment of life matter ... life processes, and shaping further layers of processing hidden behind the material image of a "finite" atom, cosmos, quantum of light, cells, at . ..

To be continued by algorithmization of control and production of the pentagon of triangles of the processing of the matter of life.

Defensive and offensive servicing as the basis for the process bases of a given situation/ phenomenon – perception/image of "matter".

Defensive and offensive servicing as the basis for the process bases of a given situation/phenomenon – perception/image of "matter", i.e., constant servicing maintains a given existential, developmental level of anything, anyone, whatever you call it.

That is, any actions can maintain, degrade or develop the basis of servicing a given state, phenomena including the property/ properties of building materials of life, photons of light, and other phenomena and processes ... breaking through the following barriers seem constant and unchanging as a system of relations in an unknown to us cosmos, but which here can be infinitely manipulated/= worked out, earned, changing its structure and properties – this applies to all phenomena, processes in the field of science, economy, politics, religion, and personal.

The issue of a proper, responsible, or holistic, global approach to matters, systems that are not suspended but interact in the surrounding dynamic global environment from the inside and outside.

What does it mean?

There may be equations 2=2+4 and E=mc2 that are ok (not in every case...), but they don't have to be. All data related to these and other models do not have to be so close if we change their service, infrastructural application, and environment; it may seem like a constant relationship, but not, ... we can also change the environment of cells, atoms, stars, black holes, quasars, galaxies... just like we changed the beds of flooding

rivers, just like we broke the barriers of the speed of sound, the first, second cosmic speed, etc., and which did not so easily enter the models of fuel quantity vs. speed, service vs. result.

This is a micro-macro universe that we can use through creative factors and participation in its hyper-evolution of phenomena and properties, which do not so easily fall within the framework of existing models and patterns. The issue of setting, setting, economy, production, which this process continues to service, – we also serve – the question of how responsibly, i.e., how much without shifting the blame and duties onto god and nature and .... stories.

It was like that with mastering the wind, water, and coal, and it also happens "by itself" in nature, but also with further changes in the speed of ... wind and the efficiency of raw materials.

This overthrows at any perceptual level all models, physical, biological, mathematical, and even your own

the old can be accepted, but at what level of usability, infrastructure / hidden structure, created, to be created ..., such a mechanical breaking of existing systems, models ... including the application of political models to states

e.g., the wall of the Ukraine-Russia problem, surrounding it with teams, and creating different international, economic, and tax relations so that it would not be profitable to fight for pieces of land when every citizen/shareholder is able to choose between paying taxes, moving, building, training, work....without a dogmatic model embarrassment.

It's about breaking the models and their environment into the first parts so that you can transform them more freely and evolve towards a more stable and developmental, not degrading development in physics, biology, ... the genome of economics, this politics, etc.

Dismantling these triangles inward and outward with reversible neo-infrastructural effects

yes, adding these cyclical triangles of decisions and infrastructure, and maybe pentagons -> shifting the boundaries of responsibility and power to the next level, patterns of triangle systems, subsequent units, including personal responsibility in this case in international, moral, economic relations ...

The exploitation of the natural environment carbon, uranium, light... influencing the change of technology, organization, economy –

raw materials1 ->technology1->raw materials 2->technology 2->raw materials n->technology n,

including evolution, economic, civilizational, and ecological development on a micro-macro-cosmic scale

EvmDNA =m1 < Eskm1DAA =m2 <Eskm2DAA <Eskm3DAA...EskmnDAA.

It's not about sacrificial evolutionary degradative in behavior, but the artistry of creation, responsibility, evolutionary explosion, progressive development, entering this mechanism of life, not matter!

It's about the efficiencies of the procedure of this life process, not the matter!

It's about a mathematical, physical, biological model that any state can be bypassed around, often literally on foot, if necessary, to solve any problem that seems to be unsolvable as material for further processing, processing room, any processing and interpretation from/on other levels of raw materials, sources, ... consciousness, perception, economic, ecological, scientific, political, social.

To be continued by share/economics (economics of exchange with nature) of nature, physics, biology, i.e., feedback on further levels, spaces of co-processing of cycles as the

basis for control, production from wild

matter, as a genuinely infinite potential for further transformations, existence, and development, as an anti-degradation approach to phenomena that are serviced or not.

CHAPTER FIFTY-ONE

When matter, time, and nature become even more flexible than we think, giving us a more extraordinary -infinite- potential.

When matter, time, and nature become even more flexible than we think, giving us a more extraordinary -infinite- potential/sources of development as well as... degradation – and it depends on how we will responsibly, effectively/productively service this potential.

Continuation of scientific, or maybe science fiction, or perhaps fiction, observations, because not including today's not very flexible views on the potential, somehow the finite potential of the nature of matter, or rather already the process of nature, which is visible in the old/modern so-called scientific, perceptual, and even expertly confirmed models of material images of nature, even a completely scientific (self-critical) approach to phenomena and our interaction with them.

What is, or rather who is, this fixative?

Fixation or fantasy, i.e., yes or no in existing models, patterns – imagining but at the same time planning, total responsibility for the shape, further shaping trends, introducing own technology, a vision of ideas, as if from the outside of nature determining directions, properties of infrastructure ... no longer nature but just the environment.

Such non-scientific leaky, possessive, and at the same time trying to determine the price, the work of this

process, this economy.

It is no longer an ordinary economy or science but rather a policy of determining the strategy of the neo-infrastructural structure of the environment to/from inside and outside, which is to be systematic, accelerating, and autonomous – we have, is to determine more or less democratically the costs of these works, this undertaking, organization, and all this above current trends, predictions, and experiences.

And that's what these articles are about, searching, suggesting the ever-deeper outline of the possibility of this neo-creation of matter and life, and whose potential is hidden in an infinite combination of processes = limitless potential of matter (determining the matter that ... does not exist), which in an endless degree affect, are supposed to affect properties today it seems to be permanent because based on dogmatic principles of seeing this process of seeing and perception is experienced .. and this is not enough to get out of this circle of environment/limited potential in which we are...

What is implied here is a deductive hyperevolutionary, hyper creational initiative (more remarkable than any force/potential of the known nature of the entire cosmos and ...more, ...and god)

as if from the inside towards the micro-macro space of imposing modeling, a means of modeling new growths of technological and physical properties...

This is deducing the school of life through play, fantasies, experiences, experiences of learning, and, above all, building a new one in an infinitely changing environment, in which one must constantly be very flexible about its negative/positive potential, and above all the infinite potential of infrastructure, properties of matter, or rather already processing/servicing... and that's why the science of physics and others can't pass their gaze stopped on the wall of the classic material approach to construction, the development of micro-macro-cosmos "matter" in every area of life, no more material but the service of our life our future.

Life in the physical, ... economic, political, civilizational...

serving, processing so that the servicing of life, the processing of the "matter" of life is as good as possible, breaking through the barriers of matter processed according to the approach to this phenomenon, being serviced, developing further the most effective infrastructural basis for development, i.e., existence, because this development is the basis of this matter/existence of any activity physical, biological, material, spiritual, civilizational.

Such not scientific but school discovery, creative learning, developing imagination and responsibility strategies in us from the inside, and not looking for means / ultimate keys outside.

This key lies in us, in our initiative, in the airtight fantasy of shaping against trends, of course, primarily, and even only for this reason, taking into account the responsible management of costs and profits in an environment of infinite potential in both positive and negative terms. An endless, unlimited, powerful approach because we are constantly dealing with the dynamics of equally infinite robust systems, inside and outside of us, which are our competitors, partners, partners ... which can never be treated as some kind of clump of stones/matter ...

The current material view of the environment leads to mental inertia in the political, social, economic, scientific, and cognitive fields.

Every phenomenon...mass/matter is the result of processing/servicing with infinite contributions of sources and potential from our and other sides of the environment inside and outside.

... entering the so-called internal/external environment of broadly understood mass/matter... destiny that determines – but we can also leave this tired shot with the infinite potential of our processing/servicing in interaction with this process matter from the outside.

Emc2 and other "constants" ... affect the behavior of the environment ... model of the environment ... but further processing, servicing, however, still determines/changes its infrastructure-> properties, its evolution ... evolution of the cosmos to the degree or above the degree of its trends...

We are also talking about the biological cosmos... within us and beyond, and its consequent infinite potential.

And this is how this imaginary or so-called real/natural wall of the environment – a clump of matter – is to be treated as a service, a process in which we take part whether we want to or not to participate in the rest – a matter of free participation in this process of physical, biological and economic changes on any level, including social, civilizational, political, national and international.

....a tax system that is a virtual share potential, literally infinitely physical! potentially, the right to participate in these or other projects is to make it easier on the national, international, and cosmic levels to influence the directions instincts of ruthless shaping, regardless of the existing balance of power, because our potential is the right determinant and not the already past balance of power.

To be continued in the strategy of subordinating the seemingly insufficiently flexible matter of problems and phenomena through processing, servicing this matter, i.e., its dematerialization, it's... servicing ... constant process decomposition into prime factors, because it is a process, or rather an infinite set of EvmDNAn processes <EskmDAAn.

CHAPTER FIFTY-TWO

Everything is based on processes with infinite potential properties, regardless of size and quantity. The issue relates to direction control in this dynamic and not a static structure, construction, causality of existence, development, agency, reason, or effect of any phenomena at the physical, biological, chemical, economic, political, social, and personal level.

The greater the interference, expectations, real expectations based on action and not only wishful thinking, not based on transferring the costs of these wishes to external natural, divine, economic factors, etc., the greater the actual responsibility, but also the potential, infinite potential, and the rights behind it, including more fundamental laws of physics, the determinants of which are the beginnings of actions bearing on those also involuntarily responsible, because they cause effects, whether we like it or not, by our direct or indirect actions (or lack of activities on our part – although they still occur fueling the effects of this neutrality, or rather garbageness), resulting from it.

And this is the fundamental law of the physics of the life matter of the micro-macro cosmos and the resulting generational consequences of creation/evolution on a scale exceeding any scales considered so far, thus causing any actions and their effects. Amen.

Because only this over-scale system of action gives any chance to get out of this specific physical, biological, and mental gravity (cage).

The objective process "construction" of "matter" -> any potential arrangements that change any institution and its properties of any object and phenomenon to an infinite degree, changes the image of a given phenomenon, i.e., the image of the matter of the phenomenon, i.e., "matter", today simply called matter in terms: ...does matter or doesn't matter of any phenomenon.

Such an operational/functional/process image of matter from stone, iron, coal, and uranium, their application, strategy planning, and laws in physical, biological, economic terms ... which intertwine, e.g., economics into physics, etc. in further interpretations and creations.. .dark matter, etc.

This description of the foundations of phenomena in the entire universe, including us, described above is a contradiction to the trash/=primitive material treatment of the world/universe and oneself in thought, in scientific and economic models.../.

it completely changes the description of the world's history, present, and future, ecology/economics, politics/religion...

Leaving this circle of current thinking will be a proper civilizational evolutionary/creative leap on the scale of humanity and literally the entire environment to an infinite extent

(by the way, we can already change our environment in the blink of an eye ... only in a negative view so far ... atomic)

-> this -> specific, global interaction at the economic, micro-macro economic level

-> proper coupling ... small models ..... science, research ... neo-environmental production.

It is supposed to be a new Neolithic revolution, but not literally on the scale of the earth, but on the cosmic scale of further, deeper spaces,

which we will not treat further as some points of a nomadic approach to exploring the micro macrocosm, but its cultivation, production, and replacement.

The transition from a nomadic, inefficient junk to a systematic moderator of world coexistence, an explosion of development, which is the basis of existence, as a negation of stability, which always is and will be a derivative of degradation and collapse.

The minor matter, and actually the smallest basic process unit, over components for a given phenomenon of particles, as the basis, and other process units soaked in the material structures of the so-called natural and so-called artificial (economic, scientific) outlets, which are further deeper stages of process levels generally included in EvmDNA in < Eskm1DAA > EvmDNA out,

where processes overlap with Esk... download

from Ev in -> and then back to Ev out

with this (further) flint knapping of the stone of science/economics -> this process as the basis for responsible science and further processes m-> Ev

where Esk always includes "m" internally and externally Ev

-m as an external expression of further process infrastructural potentials

Esk mDAA > <-Ev – m -DNA -> < EskmDAA

interventionism in this triangular system is more economical than evolutionary

divinity in it .. supported by the impact of responsibility

-> reaching divinity/religiousness ->= responsibility means moving from passive thinking to active -> listening to the system, co-creating the system

Grasping this triangular system of cycles will allow you to better proceed to effectuate the evolution of a more mathematically complete physics, chemistry, ...religion, medical processing to further raise the level of perception and influence "matter" – it is no longer matter, but process cycles, ... yes as the earth was not the center of the cosmos ... then it is not around matter but around processes that the factors of the universe of biology, physics ... are shaped ... in which we participate, and which generally means the real infinite potential from our and other sides in shaping any values properties of all phenomena, with increasing power depending on the setting of the pace/algorithm of these transformations, the production of transformations/mutations/conversions.. this production already exists, but not always with our participation, perception and awareness, or responsibility ....

that is, for example, Emc2 can have infinite connotations process internal and external, causing the so-called energy from a gram of mass of coal, uranium, etc. of matter of the same object, but in a different internal processing, can be extracted in a different/ separate/additional power of efficiency (we are also talking about a power that practically infinitely exceeds the hitherto known maximum energy efficiency), not limited by any hitherto perceptual, experimental, production, economic or mental system. Of course, we are also talking about infinite potential here in any other physical, biological ... economic parameters and their application ...

...process cycles are not always fully efficient

p-> but for the rest, they increase the perceptual potential – needed to control processes in the degree of geometric progress – infinite

To be continued in open divagations, explaining to myself and others this ... infinite perspective of creating the matter of life.

CHAPTER FIFTY-THREE

Constantly coming out of old structures and creating new structures, superstructures...

Constantly coming out of old structures and creating new structures, superstructures, and substructures is the basis of existence, functioning, life, ... not matter anymore.

We continue with the process connotations of a dynamic environment/infrastructure of matter, no longer matter.

Sometimes it is a cluster of words, ideas, suggestions, provocations, and sometimes definitions and theses, which are the basis for further application in the field of every human activity – economy, medicine, space science, quantum physics, politics...

The contradiction of two separate languages that influences the way of thinking and acting, which are intangible factors in the evolution of the infrastructural environment of civilization.

Speaking literally or not talking about the system of responsible submission/degradation or responsible exploitation/imposing, which go beyond the environment or into this environment without disturbing the existing infrastructure of the environment, using deeper autonomous process resources (independent of the material limited image of this environment) that exist.

The question of sharpness of language in the dynamic description of the environment, the problem, the "matter" of the problem.

Coming out of the gravitational collapse that exceeds the changes in or above or the side of the "m" of the infrastructural structural external and internal environment and at the same time DNA instrumental m very dynamic services/processes

and on the occasion of them changes in the direction of expansion, explosion, economy, ecology, which further affect the higher levels of m -> DAA -> n ... in the literally micro and macrocosmic degree.

-> not adapting to the nature of the environment, but adapting it to us according to proper economics.... evmdna<eskmdaaa....

These triangular EskmDAAn models... are realized in nature by various systems...including our human system – according to us.

It's a matter of being aware of this .... application in terms of discovering, the application of these elements of the triangle hidden in this triangle grid – how the laws of physics their further development and interpretation and application, etc.

Finding yourself in this jungle of threesomes.

These triangles are the actual keys (DNA-.DAA) to its management

transformation according to our needs

such competition with other systems .. for these resources of triangles surpasses existing classifications, including Menedeleev's classification.

It is about exploration, exploitation, expansion, import, but also expansive, defensive, conquering export of this in this truly infinite process potential for development and threats of the world of triangles, i.e., EskmDAAn>EvmDNAn, in which there are further these triangular systems that should be constantly identified, developed, improve...algorithmize, -accelerate this process autonomously and controllably.

Esk has two assumptions ...a-scientific, and-religious

it is a supra-material/super-divine strategy

action process supernatural, supra-evolutionary/economic classic, supra-ecological

-and the opposite of religion-science is state affirmation; it is "to fill the spirit, the void

space macroeconomics

It is .... so above religious above affirmative self-development of strategies -> degree of coverage above the environment

Esk (m1= science, religion, nature= EvmDNA)DAA

Hyperscientific, hyper-economic, hyperevolutionary consciousness

... shares of the network ... of external neurons of internal ...

...pentagon over-economy, over-science of living space

specific neo-religious nature of science/economics

pattern, hyper-scientific model entering a different area of process algorithmization...

algorithmization room for algorithmizing the processes of the environment, or rather imposing our own, or combining ours with the algorithmic superstructure of the rest of the environment, which we want to

manipulate because we are part of it, but already more and more active, conscious part, or rather part of the process, already more aware of our participation.

attitude towards .. if you have it constantly in the back of your head – strategy (pentagon ...) prepares you for war as a guarantee of the uncertainty of development as the basis of existence, its durability, maturity of existence

based precisely on the constant development/evolution of the infrastructure as such – and the process infrastructure, service infrastructure, very dynamic, autonomously developing, growing, which at first glance may have a meaningless character ... hyper natural hyper ecological, it is more than ecology / = vegetation, but taking in one's own hands its level surpassing the existing structures, ... supra-natural, ... so that the emerging new DNA of the expected processing/servicing system would protect us more effectively against severe threats from the infinite nature of the micro and macro systems of the "material" and biological cosmos. ..

to protect our own environment against threats, phenomena of further environments/settings, possibly to cooperate with them to increase our own triangular platform EskmDAA->pentagon/=defensively super-scientific> over vanity science, vanity economy

To be continued in the structuring race of the processing room...of matter and(because/of/for)life.

CHAPTER FIFTY-FOUR

Towards the growth of hyper evolutionary/economic/scientific awareness when improving the economics of the universal process integrated circuit.

Towards the growth of hyper evolutionary/economic/scientific awareness when improving the economics of the universal process integrated circuit as the foundation of evolution, production, improvement of matter, life phenomena, which are to be a new basis, an argument for expansion or degradation in every sense of the word in terms of civilization, evolution, creative.

We continue these .... against nature – as people call it –

against the laws of nature – against the often ruthless laws of nature, which supposedly must be, that it must be .... No!

Mercy and reason, not blind mental submission...scientific, economic, are the real sign of the divine,

real humanistic, actually more human, as the foundation of the power of consciousness above the so-called laws and systems, relations based in reality simply on primitive limited belief/knowledge, biasing image, plan, a strategy of action, thinking.

Above the law ... we define ... further broader, more profound, less passive legal bases – rights/laws above the existing ones according to ... charity/humanism, not scientific/economic animalization/petrifaction... We need to

be above this law of "nature".

For this, the need for further education of this flint/stone, further specialization of tools, materials, and products in a more consciously controlled phenomenal approach – as we are historically going through – such .... a neo-infrastructural process integrated circuit

neo evolutionism of the so-called material basic bricks for transmitting, that as a process as the primary use, discovery, and creation of matter, here and at a distance also in the space-time system .... retrograde, ... just any manipulation, co-creation of the neo / retro infrastructure of the matter of life. ..

This flint, this carved piece of stone, this part, this brick, tool, this so-called prime factor in my view, process – not material ... prime factors – will never be prime, but the smaller the bricks / discovered or artificial components, the it is better to repair it back, to create new ones, as other systems do with more or less, or not yet known building blocks of a given goal for each of us, each of the other systems of matter/life.

For this, you need an appropriate application, infrastructural transmission to overcome problems / avoid obstacles in the infinite potential of processing any phenomenon, from achieving supernatural goals

through the industrial/scientific boom, its mass support, artificial, deliberate, and biased.

For this freedom of life, i.e., for life because it itself is the basis of this freedom, you have to fight through this constant neo-infrastructure constantly... pushing micro and macro systems in the cosmos with these hyper investments/systems with and against other systems.

That is, about life and other material and non-material values... in processing rooms...

This supernatural processing room is, of course, planetarian engineering. For example, Denmark wants to reduce carbon dioxide not only passively but simply by storing it, pumping it to the bottom of the North Sea, surpassing the previous carbon dioxide production from Denmark by not 100 but 110% and more, as a hypernatural investment – it's just conscious planetary engineering ... and unconscious passive pollution or some kind of adaptation to...degradation.

Hyper naturally, hyperevolutionary for the infrastructural impact consistently higher than the natural background to ... in this nature of systems with often mutually contradictory tendencies, survive, be better than the parameters of these systems.

Constantly stimulate these and other hypernatural initiatives for increasing potential, chance, and development-survival direction, or else degradation will continue in the always dynamic changing conglomerate of systems inside and outside us.

....Increase in taxes, decrease in interest rates, increase in support, allowances for all infrastructural investments.

The same system of support for directions in the dynamics of infrastructural research....

This system of public-private hyper-defensive expansive investment, hyper-economics, and hyper-science is the proper stance in the war of the worlds of systems, whether we like it or not.

Arms/defense/attack improvement research is neither scientific vanity nor economic vanity because it makes hyper ecological, hyper evolutionary sense.

...These don't have to be investments of scientific or economic sense, but ->..."meaningless" supernatural

investments for /... protection of this nature – because it itself will not become a sufficiently effective armor for us against the temptations of a specific monster the nature of the external and internal before the entropic approach, the collapse of the environment in which we live ... hyper infrastructural.

External, internal infrastructural problem approach to the openness of borders, economic activity, tax reliefs, such an artificial algorithm to control economic resources and environmental potentials not based on the game, but rather on building further platforms for the areas of this ... game, nature, ... to increase the competence of responsibility in a higher level of intervention of the existence of security in this nature, above nature, being wiser than stone.

The safer, or also more powerful, infrastructure/buffer as a guarantee of further existence, as a fuller and more effective development, i.e., the more powerful the infrastructure inside and outside, the greater the chances of any action of any matter processing...not matter processing.

The doesn't matter means that any problem or issue is to be solved by action/process/service– not walking into the wall but checking to exit them around.

The conclusion is that the solution is always found outside the given structure.

Open, over-frame discussions – speak out – i.e., more comprehensive, i.e., hyper infrastructural, hyper-environmental consideration of systems as a guarantor of a safer life, there is a guarantee of effectiveness for an increasingly bold spectrum of ventures, any activities ... as a basis ... no longer matter, but safe matter – what is the basis for digressions about the meaning of life, existence, development ... which are this matter – not classical anymore, it is only a dangerous illusion ... in the long run, if we want us and our nature to live ... no longer focused on some a constant attitude of the so-called material, but a process, service, dynamic, not static attitude...

To be continued in determining and cooperating with the algorithmic infrastructure of our and other systems of nature in a very interdisciplinary approach.

CHAPTER FIFTY-FIVE

These triangular models of processing, the creation of matter, based on the processing of previous processes, which processed previous processes/approaches/procedures, which processed, process previous and future processes, ... are implemented in this nature of systems, based on this infinite grid of process rooms, which can be freely transformed into forward and backward, in depth and distance, the question of prying into the economy, the production of this process, trade, i.e., exchange, cooperation, competition, simply co-creation of nature to an infinite extent, and only depending on our commitment, not expectations ... not research structures of nature as nature, but only from the point of view of contribution, exchange, as a competitor, a competitor to be drawn in with the corresponding benefit for him(other systems of the nature/of the "matter") and therefore for us.

It is a matter of awareness and willingness to be an active creator of this structural, infrastructural reality, of

replacing, capturing, discovering, and using these elements of the triangle hidden in this infinite network of triangles – Esk(m1=EvmDNA)DAA>EvmDNA in which more or less consistent models of Newton, Einstein, etc. are hidden.

Finding ourselves in this thicket of process coupling triangles – these triangles are the keys to managing, transforming according to our infrastructural needs, securing our freedom of development, i.e., securing the environment for our life, also a life of this environment/infrastructure at the appropriate level in the dynamic cosmic world of competing systems/processes/triangles, dynamic obstacles/threats, but at the same time potentially infinite opportunities.

Such competition with other systems/processes, which are the basis of the so-called matter with which we deal, for the exploitation, expansion of this infinite potential and dangers of the world of triangles, i.e., our EskmDAAn triangles (including active developing climatic activities – planetary engineering) > other EvmDNAn process triangles, including our vain actions – evolutionarily vain EvmDNAn (e.g., climate passive or actively degrading).

Esk – self-efficiency economic has a-natural by assumption, but not against. It is a supra-material strategy (extra-natural/extra-evolutionary, super-divine), a very intense macro-economics of the micro/macro of the cosmos, as an action, a super-ecological process, above-economic classical, because modern science/religion is just a vain affirmation of the existing state.

Therefore, it is an over-affirmative, self-development strategy of using the familiarity with the environment or literally treating him as a client, partner Eskm1DAA = (Esk = our hyper-evolutionary economy)(m1= nature, religion = EvmDNA)(DAA= processing room – dynamic administrative activation).

Above the environment, i.e., above the material, or rather a process, active, potentially infinite dynamics system of phenomena hidden behind the so-called material external image, perception, of a given set of processes/functions/techniques favored in a given object of our interests, range of sight ...

It can be concluded that here, there, now, and later there will always be a solution beyond the established material – now we can better understand the phrase "it doesn't matter" – a given process, a given shot, a given case presented, a problem that could be solved by security, an attack on the so-called given wall of insurmountable problems, always alternatively omitting it, rebuilding it, opening the infinite potential of the bypass Esk...solutions.

That is, by going deeper into the matter of a given problem, the goal, or rather the processing of this matter of the problem, i.e., the processing of the process bases of this matter, i.e., the image of processing phenomena from the inside and outside, in discovering, or rather own initiatives in creating further levels, layers of these processes, new processes, so a new image of these processes, i.e., new matter ... – Esk..->Ev-> + Esk, on our part, with the participation of other parties of this infinite system of this so-called natural infinite, potentially infinite with changing thus become their properties of the process mesh.

That is further algorithmization, justification, hyper-scientific, hyper-economic education – the strategy of own production, planetary engineering, engineering on a micro-macro cosmic scale, not losing oneself in the economic-scientific directionless, passive

hyperevolutionary activities. Entering a higher area, a space beyond the current strategies... standing in the place of ... a preschooler.

Esk -n < – Esk < Ev < Esk < Esk +n... for survival/engineering explosion/working out space-time in any direction...environments controlled and automatic economically, ecologically, productively, but always with our participation as an inspiration, decision maker, stimulator, creator – an active man and not a passive stone.

To be continued in this mutual pushing of interests/ infrastructure systems, including ours.

CHAPTER FIFTY-SIX

Beyond the material, i.e., beyond (micro-macro)cosmic ideas, tools, or rather infrastructural/ tool processing – processing room.

Beyond the material, i.e., beyond (micro-macro)cosmic ideas, tools, or rather infrastructural/tool processing – processing room, which are to go to an infinite extent and potential beyond the brackets of our expectations/dreams- that is, despite powerlessness and indifference.

Opening a non/calculable, non/specific model/pattern of implementation, production – neoevolution, neoeconomics, neopolitics of environmental systems, of which we are rightful decision makers in shaping, producing its properties to a definitely infinite extent...

Can we mature into a non-static process view and treatment of nature's systems, including our process? As we have grown – not only by observation but by direct application – in our time from the transition from a geocentric to a heliocentric system, etc.?

How do you go about it speeding up the process?

From the idols of the enchanted, retrocreational materialization to the active participation in shaping the evolution of the internal and external environment – the course of this process.

The process shapes the image of "matter" at a given level of perception and use, and not vice versa, i.e.; active processing affects any course of the matter phenomenon at any spatial/temporal level. The constant of processing evolves and shapes. It is an operational definition of the properties of re or neo-construction of the

object of interest. It is based on active factors, always giving an infinite pattern, a model of transformations in any direction, shape, degree, and time.

This can be used in shaping the distances and depth of structures and then their properties of objects, which are the basis for further more or less known models and expectations.

How to continuously influence the effectiveness of the system of shaping, processing – more and more precise indefinitely without any barriers and limitations ... the infrastructure built in this way will better support the economy and science, which will have a better tool – materials for the processual recognition of the environment, its structure. In return, will it be better to support this race?

Answer.

Hypercosmic view of the phenomena of life and ... no longer matter, i.e., coming out of the material crust/vault, currently not always perceptible, although effectively determining, as it happened when going from the geocentric to the heliocentric system, like from Newton to Einstein, but here it is about something much more, no longer observation, adjustment, but taking over the process initiative over other processes – no longer the matter of the cosmos – no longer this shell of illusion, experience, observation, and civilizational, scientific, economic, social, moral, political and religious activation.

Constant serving, processing at the highest possible but "lowest" level – the so-called first parts -> guarantees the more profound, further evolutionary ... energetic level! $E/=Ev$ ordinary, existential, neo-construction defensive, offensive as a superstructure, interaction, intervention in the environment to maintain it.. develop, secure.

It is not only exploration and exploitation, but also own creation of the so-called first parts to be serviced on an increasingly profound level, and this is a ...... level to be on an increasingly deeper level, but also an autonomous level of algorithmization -> autonomy/automaticity/robotization

sending, dispatching automatic machines to create institutions/infrastructures, facilities increasingly developed inside and outside the facilities for more and more efficient........ in physical, biological, medical, cosmonautical, economic intentions, from small to large enterprises.

More and more holistic (above geo-, helio-, ...-centric) approach to the relations of "matter" processes-> outside and inside-> it is a strategy of holistic, expansive as well as creating own models of infrastructure and ... laws /->obligations-> in this macroeconomics/macropolitics of the micro-macro cosmos.

Processing EskmDAA or rather Eskm1DAA, where $m1 = EvmDNA$, processing picture of m/matter ...of space ...time for future, present, but also past of any structure processing for the history and the future, as a fundament of matter of any evolution, any creation, which are just "m" in the following, present, further level, stage of the processing room.

In this view of the matter, phenomena of matter, in some time segment, which is a very primitive counter and nothing more!...

Time – or rather some more or less primitive measure of time – which determines the course of processes forward – but (not) backward – ... but the processes themselves can be freely processed onwards and backward to the previous or future process-room process variable ..- and the timer will only determine this process .. somehow primitively but never control it! Time does not control/ maintain! But processing in its image – e.g., a car timer will not accelerate or slow down the car – you need to use a set of process rooms for this...

... in the fight against cancer, the effects of aging – reversing the disease of the process team/cluster/system/cell... – a process is a tool of process room – defining time matter space...

So we all can - and we do this passively, actively, more or less consciously together - continue on the foundation of processing ... the infrastructure of space-time phenomena – no longer classical matter and life, but closer to their definition, with consequences affecting every area of human activity and other systems of the environment surrounding us from the inside and outside- the issue is our will of the participation of creation life and matter, of the universe, of nature to be better and more robust than the universe in us and beyond.